UNDERGROUND STORAGE TANK INSTALLATION & MANAGEMENT

G. MATTNEY COLE

CRC Press
Taylor & Francis Group
Boca Raton London New York

CRC Press is an imprint of the
Taylor & Francis Group, an **informa** business

First published 1992 by Lewis Publishers, INC.

Published 2022 by CRC Press
Taylor & Francis Group
6000 Broken Sound Parkway NW, Suite 300
Boca Raton, FL 33487-2742

ISBN 13: 978-0-87371-596-6 (hbk)

Visit the Taylor & Francis Web site at
http://www.taylorandfrancis.com

and the CRC Press Web site at
http://www.crcpress.com

Library of Congress Cataloging–in–Publication Data

Cole, G. Mattney.

Underground storage tank installation and management/ G. Mattney Cole.

p. cm.

Includes bibliographical references and index.

1. Petroleum products–Underground storage. 2. Oil storage tanks.

I. Title

TP692.5.C58 1992

665.5'42–dc20[91-35124]

For Jane, Michael, and Brian

G. Mattney (Matt) Cole, an Adjunct Associate Professor at the Colorado School of Mines in Golden, Colorado, received his Ph.D. in Inorganic Chemistry from Florida State University in 1972. After a Post–doctoral Fellowship at the University of Wyoming, he was on the faculty of the Department of Chemistry at the University of Georgia until 1982 at which time he moved to the Colorado School of Mines in the Department of Chemistry and Geochemistry. Since 1990 he has been increasingly involved with the Underground Petroleum Storage Tank and Environmental Assessment programs in the Office of Special Projects and Continuing Education at CSM. He is the author of some 50 papers and abstracts.

Currently, Dr. Cole organizes and presents workshops and short courses on environmental technologies for professionals in environmentally–related fields. Among these are workshops and short courses on UST technology and management, leak detection methods, site assessments, remediation of petroleum–contaminated soils, and environmental risk and property transfers. He also has organized short courses on EPA lab protocols, aboveground storage tanks, stormwater regulations, and groundwater treatment. He administers the UST Installer/Inspector License Course and Exam for the State of Colorado and frequently gives 2– to 4–day training workshops for state UST employees.

Dr. Cole is an avid outdoors person who enjoys backpacking, hiking, backcountry and downhill skiing, and climbing mountains in Utah, Colorado, and New Mexico. He lives in Lakewood, Colorado with his wife, two sons, a Golden Retriever, a Corgi, a hamster, and a fat cat.

This book is a guide to recommended practices for proper installation and management of UST systems. It is not a substitute for the recommended practices, or for installation literature available from individual manufacturers, or for the regulations of state and local authorities. Each manufacturer has its own unique requirements for installation and warranty validation. Installers and owners should consult literature from individual manufacturers and the codes and regulations of specific states and local jurisdictions.

Acknowledgments

This book evolved over a period of two years. Originally, it was an outline for the Underground Storage Tank Installer and Inspector License Program, which I administer on behalf of the State of Colorado. As the "outline" grew, it finally seemed logical to formalize the document into its present form. Along the way I've had the privilege of meeting many conscientious installers who take pride in "doing it right!" and, of course, a few who don't. The best part has been the opportunity to work with Mike Powell, State Inspector of Oils and UST Coordinator for Colorado, and Charles Schultz, Supervisor. I owe them a large Thanks!, not only for their expertise, but also for their unfailing patience and good humor.

The other driving force behind this work is the workshops and short courses offered through the Underground Petroleum Storage Tank Program, Office of Continuing Education, at the Colorado School of Mines. The original program, consisting of a four–day UST workshop, has been in place since 1988, and recently has grown into a second tier of more specialized short courses on Leak Detection, Site Remediation, and Liability in Property Transfers. I have to thank all the participants at the four–day UST Workshops, sometimes called "UST 101". Without them there would be no workshops, and, probably, no book.

I would like to acknowledge the contributions of the instructors who have willingly and patiently shared their experience and expertise:

- Charlie Erdman, formerly with Buffalo Tank, now "retired";
- Ed Nieshoff, of the Fiberglass Petroleum Tank & Pipe Institute;
- Tony Rieck, of the National Leak Prevention Association;
- Katherine Letson, formerly with EPA Region 8, now with Maddox & Associates;
- Kevin Garrity, CC Technologies;
- Ken Wilcox, of Ken Wilcox & Associates;
- John Worland, Converse Environmental Consultants Southwest;
- Scot Donato, of Industrial Compliance, Inc.;
- and Scott Winters, Director of the UST Division of the Colorado Department of Health, and Trust Fund Coordinator.

The workshops were originally supported by a grant from the Environmental Protection Agency, Region 8, through the efforts of Ms. Debra Ehlert and Ms. Teri Bahrych. I can't say enough about Teri's enthusiasm, participation, knowledge, and support which contributed so much to the success of the Workshops.

No book is ever created in a vacuum, least of all this one. This book, in a very real sense, is a family effort. Anyone who has ever tried writing for publication knows that it is nearly impossible to proof your own work with any degree of accuracy. My wife, Jane, has read and proofed the entire manuscript an incredible number of times! Quite literally, I could not have done it without her. I am grateful for her support and encouragement in this project, and for a lot of work.

And, finally, to our sons, Brian and Michael, thanks for your patience when I've been busy, for always being ready for a hike or an outing when I needed to "get away from it all," and, most of all, for being the greatest kids anyone could hope to have.

Table of Contents

Chapter I Introduction 1
The Problem 1
Subtitle I 2
UST Installations 3
Purpose 4
Using This Book 5

Chapter II EPA Regulations for Underground Petroleum Storage Tanks ..9
Legislative Overview 9
Hazardous Materials and Waste Disposal 9
Response, Compensation, and Liability 10
Petroleum Products 11
Technical Standards 11
Definitions 12
Tank Design and Fabrication 14
Storage of Methanol–Gasoline Fuels 16
Corrosion Protection 17
Leak Detection 18
Spill and Overflow Devices 19
Piping 20

Chapter III Excavation 23
Doing it Right 23
Factors to Consider 24
Location of Excavation 26
Excavation 27
Excavated Materials 29
Finishing The Excavation 30
Depth of Cover in Dry Installations 31
Depth of Cover in Wet Installations 34

Chapter IV Material Handling 37
Care in Handling Tanks 37
Material Storage 41
Preinstallation Inspection 41
Preinstallation Tank Testing: Single-Walled Tanks 42
Preinstallation Tank Testing: Double–Walled Tanks 44
Other Equipment 46

Chapter V Backfill 49
Common Causes of Tank Failures 49
Placement of Tanks in the Excavation 50
Ballasting 53
Backfill Materials: Steel or Composite Tanks 54
Backfill for Nonmetallic (FRP) tanks 54
Installations in Unstable Soils 55

Chapter VI Supports and Anchoring 57
Flotation and Anchors 57
Types of Anchorage 59
Installation of Deadman Anchors and Hold–down Pads 61
Mid-anchoring Installations 64

Chapter VII Piping and Other Equipment 67
The Rest of The System 67
Piping Layout 67
Materials 69
Trenches and Backfill 69
Installation Procedures: All Piping 71
Installation Procedures: Metallic Piping 73
Installation Procedures: Nonmetallic Piping 74
Tightness Test For Piping: Procedure 74
Vapor Recovery and Vent Lines 75
Vapor Recovery Equipment 76
Vapor Recovery and Vent Lines: Installation 77
Manways and Equipment Access Ports 79

Chapter VIII Tank Removal and Closure 81
Safety 81
Petroleum Toxicity 82
Removal Procedures 83
Disposal 86

Chapter IX Corrosion Protection 91
Design Criteria For Buried Steel Structures 91
Corrosion in Buried Steel Structures 92
Sacrificial Anode/Cathodic Protection 95
Advantages and Disadvantages 96
Installation of Cathodic Protection Systems 96
Impressed Current/Cathodic Protection 99
Advantages and Disadvantages 101
Installation of Impressed Current Systems 102

Chapter X Spill Prevention and Containment 105
Spills and Overfills 105
Spill Prevention 105
Requirements 106
Installation of Catchment Basins 107
Secondary Containment 108
Double-Walled Tanks and Piping 110
Fully–lined Excavations 111
Vent Lines and Vapor Recovery 111

Chapter XI Release Detection Methods 115
Requirements 115
Types of Leak Detection 116
Installation of Monitoring Wells 125
Pressurized Line Leak Detectors 125

References 127

Appendix A Batteries and the Corrosion Reactions of Iron 133
Understanding Corrosion Processes 133
Batteries 133
Galvanic Cells 133
Electrolysis 138
Sacrificial Anode/Cathodic Protection 140
Impressed Current/Cathodic Protection 141
Rust 142

Appendix B Compliance Dates and Deadlines 145
Compliance Dates for USTs 145

Appendix C Financial Requirements for Owners of UST Systems 151
Requirements of Subtitle I 151
Financial Requirements 152
Insurance Compliance 153

Appendix D Summary of Federal and State Programs 155
EPA Regions 155
State Programs 156

Index 163

UNDERGROUND STORAGE TANK INSTALLATION & MANAGEMENT

Chapter I Introduction

The Problem

Out of sight and, until recently, largely out of mind, there are an estimated 2,000,000 to 3,000,000 active underground petroleum storage tank systems, USTs, across the country, mostly at service stations, mostly unprotected steel tanks, mostly "Ma and Pa" operations. In addition, there is an unknown, but probably large, number of abandoned tanks still in the ground at old service stations that have been converted to dry cleaners, parking lots, or the corner deli.

Estimates vary but it is probably safe to say that from a quarter to a half of all existing USTs are leaking petroleum into the environment, contaminating soil and drinking water sources. In some cases, where gasoline vapors accumulate in basements or sewers, leaks can provide fire or explosive hazards. And virtually all existing UST systems, even sound tanks and piping, will have some contamination due to spills and overfills.

☞ "UST" normally refers to a *system* that includes tanks, piping, and equipment. In context, "UST" may refer to a tank only.

Few people worried about leaking USTs, or "LUSTs", before the summer of 1983, at which time a series of television "exposés" drew attention to the problem of leaking underground storage tanks, and the effects of gasoline contamination on drinking water. Later that year, in testimony before Congress, LUSTs were considered to be a major source of groundwater contamination in the U.S.

☞ "LUST" — all capitals — means Leaking Underground Storage Tank.

Over the past ten years, as tanks and pipes have been excavated and removed or replaced, extensive contamination has been uncovered, particularly in vulnerable urban areas. Sites range from small, confined areas at services stations to major contamination at fueling operations and railroad yards. Because few service stations installed spill containment before being required to do so, most services station sites are contaminated with years of overfill spills. And it may be only a slight exaggeration to say that railroads have spilled at least as much diesel over the years as they have used.

The environmental damage caused by LUSTs is doubly unfortunate since it is unnecessary — UST systems that are properly designed, properly installed, and properly maintained should last almost indefinitely without leaking or otherwise contaminating the environment.

Today, underground petroleum storage tanks have become a major theater of operations in the nation's war on pollution. The Environmental Protection Agency, EPA, and many states, have adopted legislation to the effect that — provided everything goes according to plan — by 1998 all of the currently active UST systems across the country will have been replaced or upgraded.

☞ See Chapter II **"EPA Requirements"** for more information on relevant legislation.

Subtitle I

Few pieces of seemingly innocent, well–intentioned legislation have had a more profound impact on so many small businesses as Subtitle I of the Resource Conservation and Recovery Act of 1984, RCRA. Passed with relatively little dissent or debate, this Act charged the administrator of the EPA with the task of regulating the installation, operation, and management of underground petroleum storage tanks. The result was the creation of the Office of Underground Storage Tanks, OUST.

☞ The list of possible acronyms using "UST" is nearly endless — There is LUST (Leaking), RUST (Regulations for), and BUST (Bureau of). Also BUSTR, CUSTR, MUST. Georgia doesn't have a GUSTO, but it should have.

The effect of Subtitle I and subsequent EPA regulations has been to bring federal environmental legislation to many small businesses for the first time, and to create one of the best acronyms ever generated by bureaucrats. By 1998 all operations using USTs will have replaced older USTs with new installations or will have gone out of business.

Critics charge that the long–term effect will be to effectively eliminate low volume service stations, especially in rural areas. To some extent that prediction is being realized, although the problem is being addressed by the National Fire Protection Association, NFPA, and various state legislatures in the form of allowing aboveground petroleum storage tanks, ASTs, in non–urban locations.

UST Installations

Another less obvious effect has been to create a boom atmosphere in UST installations. The upsurge in tank removals and installations has been good news indeed for the legitimate UST installers; the bad news is that the opportunities, like any other boom atmosphere, have attracted the "get rich quick" types with little training and less experience. Many UST system managers have discovered to their chagrin that the "low bid" is rarely the most economical bid in the long run.

As a result, many states now require UST system installers to be licensed by the state. To acquire a license the installer normally must show proof of competence by (1) passing an examination covering recommended practices, and (2) presenting some evidence of experience in UST system installation. This is a trend that should be encouraged; it discourages unqualified, inexperienced people from attempting installations, and it provides owners with some minimum indication of competence on the part of the installer.

☞ A tank is only as reliable as the people who install it.

The combination of competent installers and new designs in tanks, piping, and equipment should eventually sharply reduce the instances of petroleum contamination, particularly in urban areas. A well–designed UST system, properly installed and properly maintained, should be trouble–free and last for many years. Indeed, some tank manufacturers are offering essentially unconditional thirty–year warranties, given only that the installation follows the manufacturer's requirements.

However, proper installation and management of underground storage tank systems involves much more than simply excavating a hole, dropping a tank in, and pumping some gas. Proper installation is an essential step in insuring against premature failure of UST systems and against inadvertent releases of petroleum products into the environment. And proper installation must be followed by good maintenance.

☞ The ultimate responsibility for proper installation and operation falls to the owner.

It is necessary that owners of UST systems be familiar with recommended practices for UST installation since owners of UST systems must certify that the system was installed according to industry standards. If the system fails for any reason, owners are financially responsible for cleanup of the resulting contamination.

Purpose

A proper installation requires practical experience, sound judgment, and careful adherence to recommended practices and procedures. The purpose of this book is to provide a concise and reasonably complete guide to the installation and management of underground storage tank systems. Detailed installation procedures are included, but more than that, the intent is to explain the reasons behind recommended practices. Sound judgment is the result of understanding *WHY*, rather than routine adherence to a list of instructions.

Efficient and cost–effective operation of UST systems follows from the combination of a proper installation and sound management practices. The current boom in UST installations has resulted in a companion boom in leak detection and monitoring equipment. Without some understanding of what the equipment is required to do, and what it cannot do, it is difficult or impossible to make intelligent choices.

The References section at the end of the book provides a thorough list of sources from which material has been derived. Even a minimum collection of these documents can run to $100 or more to acquire, and, in general, they simply list requirements, not *why*.

At the same time, it should be pointed out that this book is not a substitute for any of these Recommended Practices or industry codes, rather it is intended to complement and collate information from a great many diverse sources. The information usually is generic in nature and does not refer to a particular manufacturer. Therefore, this book is not intended to be an installation manual for specific equipment, or to be a substitute for a specific manufacturer's recommendations, or a substitute for specific manufacturer's training. If you are installing an O/C or Modern Welding tank, be sure that you consult their literature.

☞ **IMPORTANT!** Be sure to consult manufacturer's recommendations and/or literature prior to installing or operating a specific piece of equipment.

The author and publisher of this book strongly recommend that installers and owners consult the manufacturer's literature before installing or using a particular piece of equipment.

Using This Book

Since most of the chapters are written as stand–alone units and may be used independently, this book may be used as a reference source for installation and/or management of UST systems. It is also suitable for use as a conventional textbook in an installer's licensing course or in a course for regulatory personnel. The regulations, requirements, and equipment described in this book are up–to–date as of the Fall, 1991, when the manuscript was completed.

Ample room is provided in the margins for user notes and comments. Also, the information in each chapter is annotated in the margins with comments and cross–referencing to additional related topics. These comments are designed to provide reminders or additional information, or to emphasize safety precautions.

☆ **This indicates safety warnings or health hazards.**

☞ This indicates additional information or reminders.

➔ This indicates related information or additional topics.

☆ A **Star** indicates a warning or safety hazard. Please pay particular attention to these warnings, both for your safety and for the safety of those around you.

☞ A **Hand** indicates a reminder or tip or extra piece of information. These reminders are important, but not as crucial as the star warnings.

➔ An **Arrow** is a cross–reference to related or additional information. This information may be referred to later.

The book divides logically into the following sections:

- The first two chapters are an introduction to underground storage tank systems, legislation, and regulations. This material may not be as exciting as reading a Tom Clancy novel; however, it is the law with which owners and installers must live.
- The next five chapters, III through VII, deal specifically with recommended practices and procedures for UST system installation. Two chapters cover excavation and material handling. There are chapters on backfill operations, supports and anchoring for wet installations, and a chapter on piping networks and installation of other equipment. These chapters provide information and recommended practices on the basic features of UST system installations.
- The next two chapters, VIII and IX, cover specialized topics. In states that license installers, installation and removal are usually covered under the same license. Clearly, if any aspect of UST construction should be licensed and carefully moni-

➔ Site remediation is an extensive topic beyond the scope of this book. For further information see *Remedial Technologies for Leaking Underground Storage Tanks (40)*, and *Petroleum Contaminated Soils (39)*, listed in the Reference Section.

tored, it is UST removal and closure. Tank removal is far and away the most hazardous operation associated with UST system operations, and a site closure without an adequate assessment can hide serious contamination or ignore potential third party damage. Therefore, Chapter VIII, "Tank Removal and Closure," is essential for doing removals safely within federal and state regulations.

- Chapter IX, "Corrosion Protection," is a necessary topic to installers, if for no other reason than they usually must pass an exam over this material in order to be licensed, and to owners, operators, managers, and anyone installing, operating, or upgrading UST systems. Corrosion protection can be treated as a "black box;" however, the options available to owners and installers are much more apparent if you understand the basics of corrosion protection. Contrary to popular belief, it is not really a difficult subject; a little time on this chapter can be beneficial later.
- The final two chapters, "Spill Prevention and Containment," and "Release Detection Methods," concern UST system management and compliance, and include overfill and spill containment, spill and leak detection, inventory monitoring and reconciliation, tank tightness tests, and related topics.

Finally, there are several appendices with useful explanatory or summary information.

- Appendix A, "Batteries, and the Corrosion Reactions of Iron," in spite of the rather formidable title, is a straightforward explanation of why buried steel structures corrode. It is included as a supplement to the information contained in Chapter IX, and should be of interest to those who wish to pursue the topic of corrosion in more depth.

- Appendix B, "EPA Compliance Dates and Deadlines," is a summary of the information contained in Chapter II. It is presented in a concise form and should be useful to those who must comply with these deadlines.
- Appendix C, "Financial Requirements for Owners and Operators," is a summary outline of some information useful to owners and operators of UST systems, but which is outside the scope of the main text. Note that many states have more stringent requirements than these so please consult local regulations.
- Appendix D, "Summary of Federal and State Programs," is likewise included for reference. At the time of this writing, many state programs are still evolving, both in terms of program content and in personnel. All states now have a UST coordinator to contact for information, some have a separate contact for remediation, and some have still another contact for LUST trust fund activities. Therefore, the summary includes only the addresses of the relevant state agencies but does not list names of individuals.

Chapter II EPA Regulations for Underground Petroleum Storage Tanks

Legislative Overview

Installation, operation and management of UST systems is a complex and challenging task. An owner/operator or an installer must stay up–to–date and adhere closely to a complicated blend of federal statutes and regulations, overlain by state and local regulations. The result is not always harmonious or consistent. But it is the law and, therefore, it is worthwhile to briefly look at a few pertinent statutes and regulations.

The federal statutes most relevant to UST operations are:

— The Resource Conservation and Recovery Act (RCRA);

— The Hazardous and Solid Waste Amendments (HASWA);

and two related statutes:

— The Comprehensive Environmental Response, Compensation, and Liability Act (CERCLA);

— Superfund Amendments and Reauthorization (SARA).

☞ UST legislation and regulations are considerably more coherent than the current state of affairs for Aboveground Storage Tanks, ASTs.

These statutes form the legal basis for the EPA regulations of UST systems, and are the common denominator for all of the U.S. and its territorial possessions. States may enact legislation and define regulations *more* stringent than the federal standards, but not less.

Hazardous Materials and Waste Disposal

The first two statutes are part of a legislative sequence that began with the **Solid Waste Disposal Act of 1965**, SWDA, the initial attempt by Congress to address the problem of municipal and sanitary landfills and waste disposal. In 1976 the Solid Waste Disposal Act was amended as part of the **Resource Conservation and Recovery Act,** or RCRA (pronounced "*Rick* rah"). RCRA dealt extensively with the problems of hazardous waste handling and disposal.

During the summer of 1983, following highly publicized exposés on *60 Minutes* and *Good Morning America,* EPA representatives presented testimony to Congress to the effect that leaking underground storage tank systems were the single greatest source of groundwater contamination in the United States. Subsequently, Congress passed a bill, entitled the **"Hazardous and Solid Waste Amendments"** (usually referred to as "HASWA"), which reauthorized and amended RCRA. The bill was signed into law in November, 1984 by President Reagan.

☞ SWDA, RCRA, and HASWA are all part of a legislative sequence of Acts and Amendments sometimes collectively called, "RCRA."

The Act is divided into parts called "Subtitles": Subtitle C deals with hazardous materials and wastes, Subtitle D covers landfills, and a major portion of this new legislation, Subtitle I, defines Underground Storage Tank Systems (USTs), provides for the development and implementation of a comprehensive regulatory program, and requires the EPA to set standards for operation of UST systems:

> The Administrator...shall promulgate release detection, prevention, and correction regulations applicable to all owners and operators of underground storage tanks, as may be necessary to protect human health and the environment.

Subtitle I also requires owners of UST systems to notify the appropriate agency of the state in which the UST system is located of the existence of their UST systems, and encourages states to implement their own UST programs. Owners of UST systems located on Indian reservations must notify the regional office of EPA.

Response, Compensation, and Liability

☞ SARA §9003(h) gives EPA and states the authority to require owners to clean up releases from petroleum UST systems.

CERCLA, also known as "The Superfund Law," was passed in 1980 in an effort to clean up and establish liability for sites contaminated with hazardous wastes; however, CERCLA affects UST system owners and operators in several ways.

In 1986 Congress reauthorized and amended CERCLA with a new law entitled, **Superfund Amendments and Reauthorization Act**, or SARA, which further affected UST system operations since SARA also amended Subtitle I of RCRA. One provision of SARA established a trust fund to help the EPA and state agencies finance cleanup activities associated with leaking underground storage tanks. The "LUST" fund was renewed in 1990 by the 101st Congress with the imposition of a 0.1 cent per gallon tax on most fuels.

☞ The "petroleum exemption" under CERCLA means compliance with regulations for petroleum fuels is less expensive relative to hazardous materials or hazardous wastes.

Petroleum Products

Petroleum products are defined in CERCLA as regulated substances rather than hazardous substances, and are specifically exempted from the provisions of RCRA, Subtitle C. What this means is that petroleum products are much less expensive to store and handle than they would be otherwise. In fact, there is a conflict in definitions between Subtitle C and Subtitle I. Benzene, toluene, xylenes, ethylbenzene, and methanol are all present in appreciable quantities in motor fuels, especially in automotive fuels, yet all are listed hazardous substances under Subtitle C.

Technical Standards

On September 23, 1988, after three years of public hearings and comments, the EPA published final regulations for new and existing USTs entitled ***Technical Standards and Corrective Action Requirements for Owners and Operators of Underground Storage***

Tank Systems (referred to hereafter as *Technical Standards* or simply the regulations).

These standards are collected in the *Code of Federal Regulations*, Volume 40, which includes all federal environmental regulations. Parts 280 through 281 refer specifically to underground petroleum storage tanks. Thus, many documents on USTs refer to 40 CFR §280–281.

Although *Technical Standards* sets requirements for all states, in practice the EPA is empowered to delegate enforcement authority to the states through one or more "implementing agencies", which may set requirements more stringent than *Technical Standards*, but not less. In Colorado, for example, the implementing agencies are the State Inspector of Oils, which has jurisdiction over installation, operation, and removals of UST systems, and the UST Division of the Department of Health, which has jurisdiction over contaminated sites and remediation. In many states yet another agency is being established to administer state LUST trust funds.

Definitions

A **regulated substance** is defined in §101(14) of the Comprehensive Environmental Response, Compensation and Liability Act (CERCLA) and includes all petroleum products such as automotive, diesel, and aviation fuels.

☞ **Regulated substances** and **hazardous substances** are not the same. A **hazardous waste** is a hazardous substance that is no longer a product.

A **hazardous substance** is defined in RCRA and CERCLA as a product that has certain characteristic or listed properties, including ignitability, corrosivity, reactivity, or toxic characteristics. Specific substances are listed in *Technical Standards*. A hazardous substance is generally acutely or chronically toxic, or is hazardous to human health or the environment. Hazardous substances include many industrial "chemicals" and commercial products, but not petroleum products and natural gas, which are explicitly excluded.

☞ Regulated substances that contain hazardous substances are a grey area in the regulations — at least at the time of writing.

It is important to note that while petroleum and petroleum products are currently excluded, the time may come when automotive fuels will not be excluded since, as noted above, many automotive fuels contain components and additives such as benzene, xylenes, and methanol, which are listed hazardous substances under CERCLA. This is an important consideration for anyone involved in tank installation, operation, removal, or remediation.

☞ In operation a UST is a *system*. Tanks cannot be used without pipes, pumps, vents, etc. Except during installation UST refers to the system.

An **Underground Storage Tank System, UST**, as defined in Subtitle I of RCRA, is a tank or tanks, used to store petroleum products, together with piping and associated equipment, that is 10% or more underground. There are nine exclusions from the definition of a UST system:

- Farm or residential tanks of 1,100 gallons or less capacity storing fuel for non-commercial purposes.
- Heating oil tanks where the heating oil is used on the premises.
- Septic tanks.
- Pipeline facilities.
- Surface impoundments, pits, ponds, and lagoons.
- Stormwater or wastewater collection systems.
- Flow–through process tanks.
- Liquid traps or lines directly related to oil or gas production or gathering operations. (This does **not** include oil or grease traps in service stations.)
- Storage tanks in an underground area, such as a basement, cellar, or underground vault. This exclusion includes storage tanks that are underground, but not in the ground.

The EPA also defines four exemptions from the requirements of *Technical Standards*:

- USTs containing hazardous wastes or mixtures of hazardous wastes and regulated substances. These are regulated under Subtitle C of RCRA.

☞ It is important for installers and owners and operators to be aware that some additives and aromatic components of automotive fuels *may* be classed as hazardous materials under EPA's TC rule.

- Machinery containing regulated substances for operational purposes, including hydraulic lift oil storage tanks.
- Wastewater treatment tanks (regulated under the Clean Water Act).
- Tanks of less than 110 gallons capacity, and other small quantity tanks. 110 gallons is smaller than the smallest commercially available UST, which is 275 gallons.

New and existing tanks are defined as follows:

- Existing petroleum storage tanks are those installed prior to December 22, 1989.
- New petroleum storage tanks are those installed on or after December 22, 1989.
- All petroleum storage tanks must be in full compliance by December 22, 1998.

☞ *De minimus*, as defined in SARA, is the minimum amount of a hazardous material that can be present at a location without having to be reported.

An **owner** is defined as: (1) in the case of a UST system in use on November 8, 1984, or brought into use after that date, any person who owns a UST system used for storage, use, or dispensing of regulated substances; and (2) any person who owns such a UST system immediately before the discontinuation of its use.

An **operator** is any person in control of, or having responsibility for, the daily operation of the UST system.

A **petroleum UST system** is an UST system containing petroleum or mixtures of petroleum with *de minimus* quantities of other regulated substances. Such systems include motor fuel, distillate fuel oils, residual fuel oils, lubricants, petroleum solvents, and used oils. At the present time the exact definition of *de minimus* is being decided by the EPA on a case–by–case basis.

Tank Design and Fabrication

EPA's final regulations require that all tanks be of one of three alternative designs:

- fiberglass-reinforced plastic, FRP;
- coated and cathodically protected steel, such as the STI-P3 design; or
- composite fiberglass- or epoxy-clad steel tanks.

Tanks must meet the standards of a "nationally-recognized" institute or laboratory, which means, in effect, tanks must meet or exceed standards set by one or more of the following:

☞ The Association of Composite Tanks (ACT) has been combined with The Steel Tank Institute; however, ACT–100 remains as an industry specification.

- Steel Tank Institute (STI);
- Fiberglass Petroleum Tank and Pipe Institute (FPTPI);
- American Society for Testing and Materials (ASTM);
- Association of Composite Tanks (ACT);
- Underwriters' Laboratory (UL); *or*
- Underwriters' Laboratories of Canada (ULC).

Single–wall fiberglass– or epoxy–coated steel composite tanks should be labeled with one of the following design specifications:

☞ "STI–P3" means Steel Tank Institute Protected 3–Ways: coated, built–in anodes, and electrically isolated.

- ACT–100 and UL 58;
- ACT–100 and UL 1746; or
- UL 1746.

Cathodically protected steel tanks should be labeled with one of the following design specifications:

- STI–P3 and UL 58;
- STI–P3 and UL 1746; or
- UL 1746.

FRP tanks should conform to UL 1316.

Installers should note that USTs with a fiberglass reinforced urethane coating, without cathodic protection, do not meet any of the EPA specifications and standards. ACT–100 requires a fiberglass reinforced isophthalic resin only and no other material may be substituted. (‡ Robert Holland, *Journal of the Association for Composite Tanks,* July–August, 1990).

As fuel additives evolve to meet stricter air quality standards, it is important to note that the EPA also requires that tanks be constructed of materials compatible with the liquid to be stored in the tank. This means that the compatibility of FRP systems and the proposed stored fuel should be carefully considered, particularly if a new fuel is to be stored in an FRP tank of unknown age and origin.

☞ Methanol, CH_3OH, is added to automotive fuels to reduce the emissions of carbon monoxide, CO.

In general, there are three types of UL-listed FRP tanks currently available:

- for petroleum products only;
- for petroleum fuels containing up to 10% ethanol;
- petroleum products containing more than 10% methanol or ethanol, up to 100% methanol.

Storage of Methanol–Gasoline Fuels

Fuels with a very high methanol content, such as M85 (85% methanol), can reasonably be expected to become increasingly common in the future. At the time of writing, EPA has just proposed a "clean fuels" regulatory program to control toxic emissions associated with gasoline use (56 FR 31176, July 9, 1991). The proposal would require use of oxygenated fuels (mostly methanol) in 41 non–attainment areas during the winter months. According to the report, "Retail service stations offering oxygenated fuel would have to establish properly labeled segregate storage for the oxygenated fuel."

☞ Reported in **Underground Storage Tank Guide**, *Bulletin,* September, 1991. See Reference 36.

Therefore, conscientious installers must be aware of regulations and potential compatibility problems, and be prepared to advise owner, operators and architects alike of requirements and options. Similarly, prudent owners should insure that new installations are prepared for future fuel blends.

Unlined steel tanks are suitable for storage of gasoline–methanol fuel blends, although older tanks should be cleaned before using. Internally lined tanks and retrofitted tank liners may or may not be suitable for methanol storage depending on which relining method is used. The manufacturer should be consulted.

If converting older tanks, make certain that the tanks are cleaned of rust and water. Water will extract methanol into an acidic aqueous solution that can be damaging to internally lined and FRP tanks.

☞ Methanol solutions will not induce corrosion in steel tanks, but mat affect soldered joints.

> **WARNING:** Methanol is highly soluble in water and forms an acidic solution that is both corrosive and toxic to human health.

Fiberglass–reinforced plastic tanks may or may not be suitable for methanol–gasoline storage. The manufacturer should be consulted.

Corrosion Protection

Underground petroleum storage tank systems are required to be protected against corrosion. For new installations corrosion protection must be in the form of one or more of the following:

☞ New steel tanks that follow the STI-P3 design specifications come with factory installed corrosion protection.

- Coating;

  and

- A cathodic protection system, either sacrificial anode or impressed current protection system;

  or

- Be constructed of a material that does not require corrosion protection.

☞ After December 22, 1998 corrosion protection will be required on *all* UST installations regardless of age.

Existing steel tanks must be retrofitted with one or more of the following:

- A cathodic protection system — either sacrificial anode or impressed current protection system; *or*
- By relining and periodic tank tightness testing; *or*
- Both.

Leak Detection

EPA regulations require some form of leak detection or release monitoring on all FRP, composite, and steel tanks by 1993, or a combination of inventory control and tank tightness testing up to 1998 depending on age and type of corrosion protection. Leak detection devices must be able to detect a leak of 0.1 gallons per hour with a probability of detection of 0.95 and a probability of false alarm of 0.05. Leak detection requirements may be satisfied by one of the following two procedures:

☞ After December 22, 1993 leak detection will be required on *all* UST installations, regardless of age.

1. Monthly monitoring by one of the following methods:
   - Automatic tank gauging; *or*
   - Vapor monitoring; *or*
   - Interstitial monitoring; *or*
   - Groundwater monitoring; *or*
   - Statistical inventory reconciliation; or
   - Other EPA–approved methods.

2. Monthly inventory control and tank tightness testing, depending on the condition of the UST system under the following circumstances:

Compliance deadlines

Installed by Dec. 22 —	Comply by Dec. 22 —
before 1965	1989
1966–69	1990
1970–74	1991
1975–79	1992
1980–88	1993

If an existing tank has corrosion protection and spill/overfill protection installed —

- Monthly inventory reconciliation; *or*
- Inventory control and tank tightness testing every 5 years; *and*
- Add monthly monitoring devices by December 22, 1998.

If an existing tank lacks corrosion protection and spill/overfill protection—

- Monthly inventory reconciliation; *or*
- Inventory control and annual tank tightness testing; *and*
- Replace or upgrade system by December 22, 1998.

3. New systems installed after December 22, 1988 or existing systems may use inventory control and tank tightness testing for a maximum of 10 years after installation or upgrade.

Owners and operators must note that in order to use vapor monitoring or groundwater monitoring to satisfy leak detection requirements, a site assessment must be performed to establish background levels and insure that contamination is not already present. Wells must be set according to regulations in the backfill zone and in permeable soil.

Spill and Overflow Devices

☞ A spill of 25 gallons or more, or a spill that produces a sheen on standing water, *must* be reported to the appropriate implementing agency.

Since the most common form of contamination around USTs results from spillage during tank fill operations, EPA regulations require spill/overfill containment on all new installations and on all installations after 1998.

All new UST systems must also be equipped with the following devices:

- catchment basins

and

- overfill protection by installing one or more of the following devices:
  — a device which will alert the fuel transfer operator when the tank is no more than 90% full by restricting the flow of fuel into the tank or by activating an alarm; *or*
  — a device that will automatically shut off fuel transfer when the tank is no more than 95% full; *or*
  — an equivalent device approved by the EPA.

All existing UST installations must be equipped with one or more of the following:

- Catchment basins equipped with manual or automatic drains;

  and
- Automatic shutoff devices *or* overfill alarms;

  or
- Ball float check valve.

Piping

All new piping must be of the following materials:

- Coated and cathodically protected steel;

  or
- Fiberglass;

  or
- A design or materials approved by the EPA, an EPA–designated third party, or a national laboratory or institute.

Compliance deadlines

Installed by Dec. 22–	Comply by Dec. 22–
before 1965	1989
1965–69	1990
1970–74	1991
1975–79	1992
1980–88	1993

By December 22, 1998, all piping, existing or otherwise, must be of the following materials:

- Coated and cathodically protected steel;

  or
- Fiberglass;

  or

- A design or materials approved by the EPA, an EPA–designated third party, or a national laboratory or institute.

☞ All pressurized piping must have been equipped with automatic flow restrictors by September 22, 1990.

All new pressurized piping must be equipped at installation with the following:

- Automatic flow restriction device;

  or
- Automatic shutoff device;

  or
- Continuous alarm system;

  and
- Annual line tightness testing or monthly monitoring.

Leak detection requirements and compliance dates for piping follow those for tanks. All existing piping must be similarly equipped by 1993, which means that most or all existing piping in place in 1988 must be in compliance or be replaced between now (1991) and 1993.

Automatic leak detectors must be capable of detecting a leak rate of 3 gallons per hour at 10 pounds per square inch (psi) for one hour with a probability of detection of 0.95 and a probability of false alarm of 0.05. The deadline for third party certification for manufacturers of line leak detectors is September 22, 1991.

➔ See Appendix B for more details on compliance deadlines.

Suction piping, which is the rule in European installations, but is less common in the U.S., has somewhat less stringent requirements. Existing suction piping must have one of the following:

- Monthly monitoring;

  or
- Line tightness testing every three years;

  or
- No requirements if the system meets the following design requirements:

- The piping slope is adequate so that product can drain back into the tank when suction is released;
- There is only one check valve which is located as close to the pump as possible; and
- The installation is verifiable as to the preceding two requirements. A statement by the owner/operator is not sufficient.

The idea behind these requirements is to provide a single path for product, and that path must lead back to the tank rather than into the environment. However, it is important to note that details of suction piping installations must be verifiable by inspectors. If the installation is backfilled, as built video tapes or photos are recommended.

Chapter III Excavation

Doing it Right

Data from EPA surveys indicate that about 25% of all leaking UST systems are caused by improper installations. Installation mistakes that can cause tanks or piping to leak include:

- Inadequate pit and trench design;
- Inadequate excavation procedures;
- Improper tank or piping depth;
- Improper handling of materials;
- Inadequate bedding and placement of tanks and piping;
- Poor or unsuitable backfill material;
- Inadequate compaction procedures;
- Inadequate or improper anchoring in high groundwater conditions; and
- Improper installation of attachments, particularly joints and fittings.

The first step in any installation is the excavation. Excavation is expensive and there is a tendency to reduce bids by skimping on the amount of excavation necessary. It is certainly true that too much is wasteful, but too little can be hazardous to worker safety and is a poor construction practice. An efficient procedure at this point is cost–effective, economical, and can prevent doing the job over later.

Typical UST installations in service stations in urban locations must be squeezed into relatively small areas, which limits the options available for excavation size, beveling excavation sides, storage of materials, and clearance for tanks and piping. The prudent installer will always take advantage of all the room possible.

Factors to Consider

Before excavation begins, the installer should consider several points:

☞ Following manufacturer's recommendations is essential to meet all the conditions of equipment warranties.

- Manufacturer's recommendations: Follow the tank manufacturer's recommendations for installation of the specific tank size, type, and location. This is an important consideration to avoid violating the conditions of the warranty. Likewise, follow manufacturers' recommendations for piping and other equipment.

☞ Special conditions such as sandy soils or old landfills may require additional clearance, up to two feet or more, to avoid migration of backfill.

- Soil type, resistivity, and stability: These factors affect the safety of workers, the size and extent of the excavation, the possible need for filter fabric, and the detailed features of corrosion protection.
- Surface water (runoff) conditions: If runoff from snow or heavy rains is a problem, it will be necessary to ballast tanks as they are placed in the ground.

☞ See Chapter VI, **"Supports and Anchoring,"** for further discussion of flotation.

- Depth of water table at its highest: The depth of the water table will determine whether supplementary anchorage is necessary.
- Space requirements: Is there adequate space away from the excavation for backfill storage if it must be stored temporarily? Similarly, is there ample room for tank and equipment storage?

☞ Always assume that an installation will be subject to traffic loading sooner or later unless conditions specifically prohibit traffic access.

- Backfilling and compaction requirements: What type of backfill is suitable for the soil type? Is the excavation large enough to allow workers and/or compactors between tanks?
- Traffic requirements: How much and what type of traffic passing over the tank installation will affect depth of cover and finish slab. Also, be sure to

allow ample room for tankers to maneuver into and out of the site.

- Space for associated equipment: Be sure there is adequate space to run vent and vapor recovery lines, electrical conduit, and monitoring equipment.
- Piping requirements: Is there ample room to run piping without crowding. This is particularly important in typical urban service station installations.

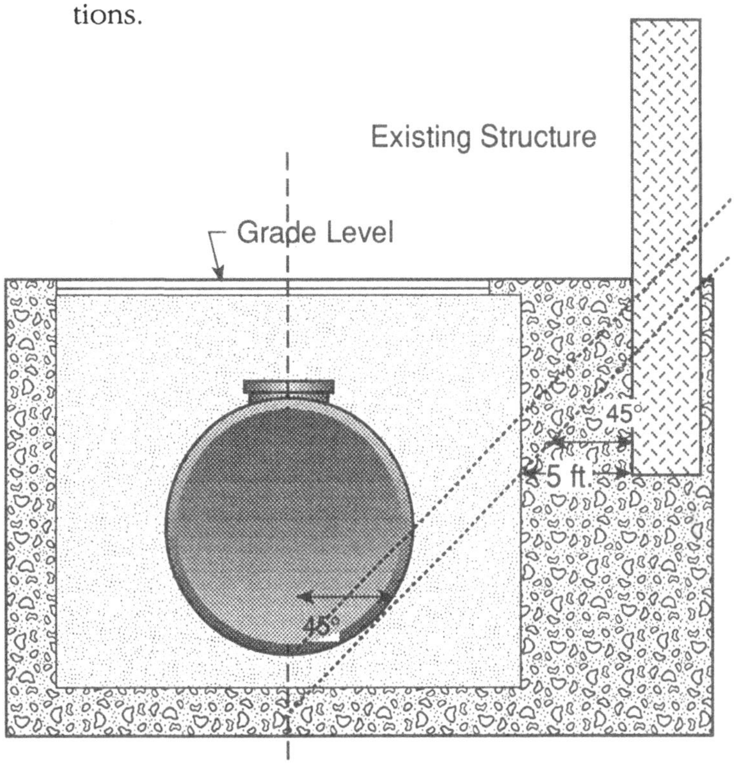

Figure III–1. The 45° Rule. *To determine the correct distance from an existing structure for placement of a tank, in the absence of local building codes, measure a minimum of 5 feet out from the bottom of the existing foundation, then a 45° line should strike the lower quadrant of the tank. Note that this rule places the centerline of a 10–foot diameter tank 10–20 feet from the nearest structure, which may not be feasible in an installation where space is limited.*

☞ Be sure to check local building codes for offsets from property lines and structures.

Location of Excavation

In general, the excavation should provide space for tanks, piping, associated equipment, and sufficient backfill to insure adequate support.

The excavation must be made with due care to take into account adjacent structures or possible future structures. Consideration should be given to the following before beginning the excavation:

- Avoid undermining foundations and supports of existing structures;
  and
- Avoid the possibility that the UST will become loadbearing *under* future structures. USTs are *not* designed to support building foundations.

In the absence of local building code specifications, a minimum of 3 to 5 feet from property lines or from existing or future foundations is recommended.

The rule illustrated in *Figure III–1* is useful to determine tank placement if space allows, and in the absence of local building codes. However, applying the rule to a 10–foot diameter tank buried 36 inches below grade places the centerline of the tank 13 feet from an existing structure, or the approximate edge of the excavation is 6 feet from the structure *(Figure III–2)*. That may be a generous amount of space in a typical service station installation.

Before excavation proceeds try to determine the location of underground utilities, such as buried conduit, water or gas lines. In many areas the exact location of underground utilities is not known with certainty. Be prepared for all possibilities and emergencies. Have a plan of action ready in the event a utility line is broken, especially a natural gas pipe or electrical conduit. Discuss the plan with workers; don't wait for the emergency to happen.

Note the location of overhead utilities. In particular, high voltage overhead power lines pose a threat to the operation of cranes and heavy equipment.

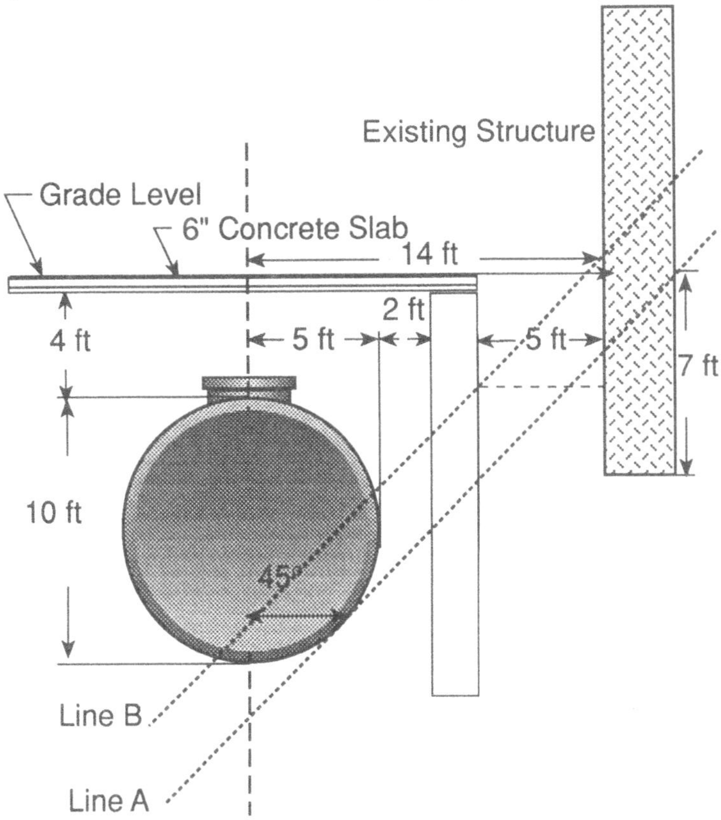

Figure III–2. Tank Placement. *Applying the 45° rule based on an existing structure extending 7 feet into the ground and 2 feet of clearance places the centerline of the tank 14 feet from the structure (Line A). If the structure extended only 4 feet down, the centerline would be approximately 16 feet from the structure (Line B). The 2–foot shaded area is an optional amount of space.*

☆ **OVERHEAD POWER LINES ARE A SERIOUS HAZARD TO WORKERS USING CRANES AND OTHER HEAVY EQUIPMENT.**

Excavation

Burial depth is the main restraint against tank buoyancy. If the burial depth is at least 60% of the tank diameter, supplemental restraint is normally not necessary. However, since the total loading on the upper load bearing surface of the tank should not exceed 5 psi, all manufacturers recommend that the maximum burial depth for tanks should not exceed the tank diameter. That is, a 4 foot diameter tank should never be buried more than 4 feet deep.

In fact, the maximum recommended burial depth for any tank is 7 feet measured from the top of the tank to the top of the finished grade.

Since adequate backfill is essential to provide lateral support for tanks, the excavation itself must be large enough for 12–24 inches clearance between adjacent tanks and between tanks and side walls.

→ See API RP 1615, §5.3. Reference 1.

- The American Petroleum Institute, API, recommended practice is a minimum of 12 inches for steel tanks or 18 inches clearance for FRP in all horizontal directions.

→ See PEI RP 100–90 §4. Reference 1.

- Petroleum Equipment Institute, PEI, recommends a minimum of 24 inches clearance for all tanks, regardless of type.
- 18 inches of clearance all around is usually sufficient for most normal installations, in which the native soil is stable and compact.

In addition all tanks require a minimum 12-inch bedding layer.

☆ **NATIVE SOIL OR BACKFILL PILED NEXT TO THE EXCAVATION CAN BE A SERIOUS SAFETY HAZARD.**

All excavations must be sloped or shored in accordance with OSHA regulations. If necessary, a soil engineer should be consulted to determine the degree of slope required to stabilize the sides.

> **Warning:** Since workers must enter the excavation at some point in the installation, it is important to insure that the excavation walls are safe from slumps or slides.

If space is limited and the sides cannot be sloped, shoring should be used *(Figure III–3).*

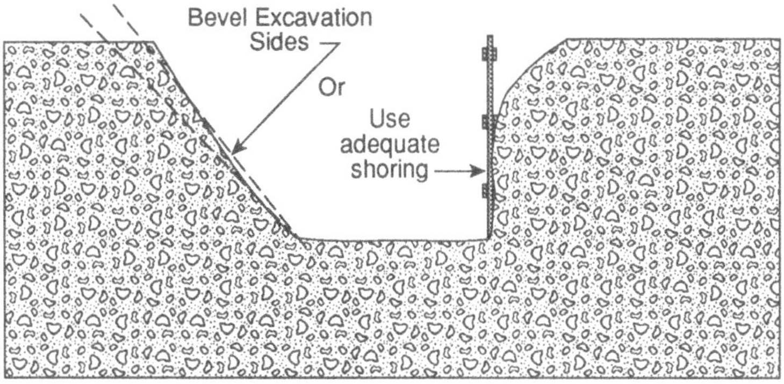

Figure III–3. Excavation Safety. *Excavations must be adequately shored or beveled past the angle of repose of the native soil. Vertical sides are acceptable only in clays or solid rock. Since typical service station installations have limited space, temporary shoring may be the best alternative to beveling.*

☞ Excavated native soils are **NEVER** suitable as backfill materials.

Excavated Materials

Excavated materials are never suitable as backfill since native soil is almost uniformly inhomogeneous and may contain cobbles and/or other debris. Therefore, all material excavated from a hole should be removed from the site as the excavating proceeds, if possible. Removing excavated materials is not only safe practice it is also cost–effective since the material must be handled only once.

Excavated materials that are not removed immediately should be piled as far from the hole as possible to avoid the possibility of loose material falling back into the hole creating a hazard to both workers and excavation. In the absence of local building code specifications, a minimum of 3 to 5 feet from property lines or from existing or future foundations is recommended *(Figure III–4).*

Ideally, backfill should be used as it is delivered to the site. If backfill must be stored temporarily, place the pile well back from the excavation in a clean area, free of debris.

☞ The bottom of the excavation must be level, smooth, and free of debris.

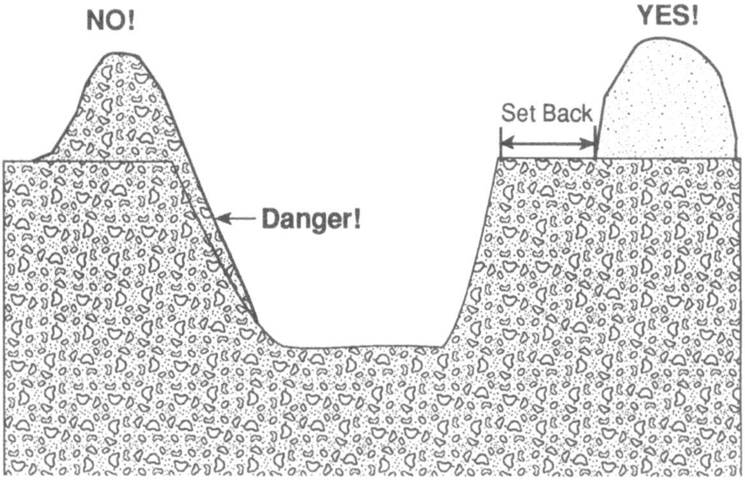

Figure III–4. Proper Handling of Excavated Materials. *If excavated soils cannot be removed from the site immediately, they should be stored well away from the edge of the excavation. Similarly, backfill materials, if not used immediately, should be stored well back from the edge of the excavation.*

Finishing The Excavation

Among the major causes of tank damage in new installations are the following:

- Setting the tank directly on native soil without removing rocks and debris;

  and

- Improper bedding that contains debris, such as rocks or cobbles;

  or

- Bedding that has not been properly compacted and contains voids.

Therefore, before the bedding layer is placed, check the excavation carefully for rocks or debris. Be sure the bottom is level. If anchors, such as hold–down pads or deadmen, are to be used, they should be installed at this time.

If tanks with bottom sumps are being installed, the excavation bottom should be modified to provide a 12–inch deep and 24–inch diameter hole centered on the sump location.

Filter fabric should be used between backfill and unstable soils to prevent backfill migration. Fine grained native soil, such as sand, can migrate through the interstitial spaces in pea gravel or crushed rock. The backfill can in turn migrate to fill the spaces left in the native soil. The eventual result is formation of undesirable voids in the backfill.

In special circumstances where secondary containment is necessary, lining the excavation with geotextile cloth may be required. If a geotextile liner is used, it must wrap around to cover the entire installation. Otherwise, since, by definition, geotextile liners are impervious, the UST will be sitting in a bathtub.

Depth of Cover in Dry Installations

Burial depth is an important consideration for three reasons:

☞ NFPA 30 (¶ 30–20) recommends 36 inches minimum burial depth in traffic areas.

- Tanks must be protected from the weight of traffic;
- There must be adequate clearance for equipment; and
- Weight of the backfill over the tank is the principal restraint against flotation.

A minimum burial depth of 36 inches is usually taken to be a standard. In practice, however, burial depth is determined by a complex mixture of variables in which primary consideration must be given to piping runs and auxiliary equipment. Consideration must also be given to traffic loads and depth to the water table.

☞ 1 inch of reinforced concrete is approximately equivalent to 2 inches of compacted backfill.

A rule–of–thumb to estimate the *minimum* amount of backfill for 4– to 10–foot tanks in traffic areas is to use 36 inches as the basic burial depth and subtract 2 inches for each 1 inch of reinforced concrete used in the finish slab. Asphalt replaces backfill on approximately a one–to–one basis; that is, one inch of asphalt is approximately equivalent to one inch of backfill *(Figure III–4).*

> ***Example:*** An 8–inch reinforced concrete slab will be laid at grade level. An 8–inch slab is approximately equal to 16 inches of backfill, so the minimum amount of backfill is 36 — 16 = 20 inches of backfill.

The following table gives the combined PEI and API recommendations for minimum burial depths for 4– to 10–foot steel, composite, or FRP tanks.

☞ These are *minimum* burial depths, not recommended depths.

Depth of Cover for USTs

Type of Slab at Grade	Asphalt		Concrete		None
	Slab thickness	Backfill depth	Slab thickness	Backfill depth	Backfill depth
Traffic Areas	6–8"	30"	6–8"	18"	36"
Non–traffic Areas	4–6"	12"	4–6"	12"	24"

Note that these minimum burial depths rarely allow adequate room for piping slope or equipment installation.

A tank greater than 10 feet in diameter should be installed in accordance with manufacturer's directions. This statement certainly applies to all tanks; however, it

☞ Excessive burial depth can cause structural damage.

is particularly applicable to large diameter (over 10 feet) tanks, and especially to FRP tanks, which must be protected from excessive backfill loading.

For 12–foot diameter tanks the minimum basic burial depth is 48 inches, less an appropriate amount for a slab at grade.

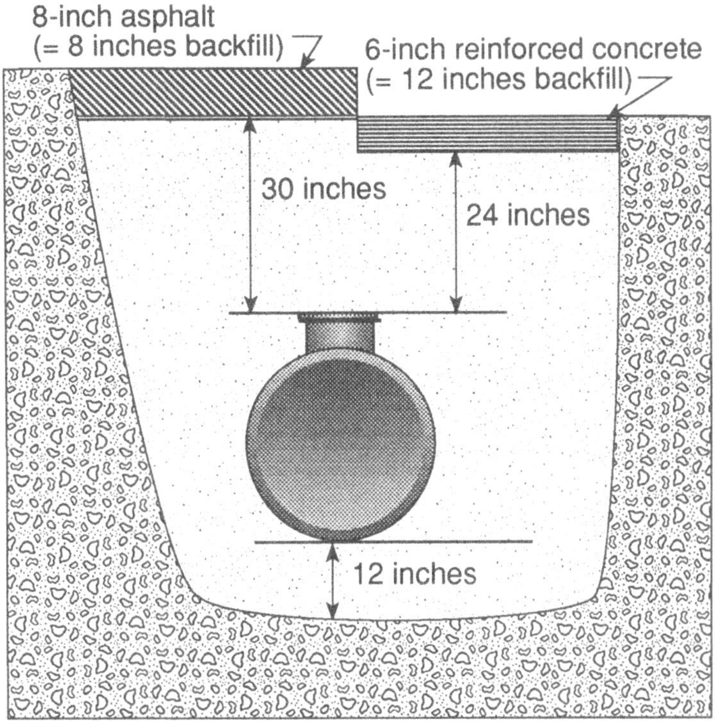

Figure III–5. Depth of Cover for Dry Installations. *Adequate depth of cover is essential to protect the tanks from traffic weight and from flotation. Consideration must be given to piping runs and added equipment, such as vapor recovery, spill basins, and monitoring devices.*

☞ The *maximum* burial depth for any diameter tank is 7 feet.

For any diameter tank where the burial depth is determined by piping slope and attached equipment, the maximum burial depth is 7 feet measured from the top of the tank. The maximum is determined by the load pressure that the backfill exerts on the top of the tank. The maximum pressure on the top of a tank is approximately 5 pounds per square inch. Excessive burial depth can exert enough pressure on the surface of the tank to cause structural damage to the tank.

After backfill has been added to the top of the tank, the tank curvature, or deflection, should be carefully checked. Deflection limits are determined by the individual manufacturer. If the deflection is within acceptable limits at this point, and if the backfill has been compacted properly, the remainder of the excavation may be safely backfilled.

Depth of Cover in Wet Installations

☞ See Chapter VI, **"Supports and Anchoring,"** for more information.

The specific gravity of typical petroleum products stored in USTs averages around 0.85, relative to 1.0 for water. Since the weight of the UST plus product is less than an equivalent weight of water, tanks will experience a positive buoyancy when water is present in the excavation. In short, tanks float *(Figure III-6).*

Since most tanks are set three to five feet deep to accommodate piping slope and auxiliary equipment, wet installations — which is any excavation where the water level may, at any time, reach the bottom of the tank — require supplemental anchorage for the tank; therefore, the above specifications do not apply *(Figure III-5).*

☞ Plan for the "hundred year" water table.

The origin of the water in the excavation may be due to the existence of a high water table, but may also be due to surface runoff (the "bathtub" effect). Surface runoff is a particular problem where an installation may be surrounded by large areas of asphalt, as in shopping centers or industrial facilities.

In areas as diverse as Florida and Colorado installers must cope with high water tables whose levels fluctuate depending on the time of year. Heavy rains, snowmelt, tidal fluctuations, or proximity to surface water, such as rivers, irrigation ditches, or creeks, can cause groundwater levels to vary over relatively short periods of time. Even in areas where the nearest aquifer may be hundreds or thousands of feet deep, an excavation set in clay may flood every time it rains.

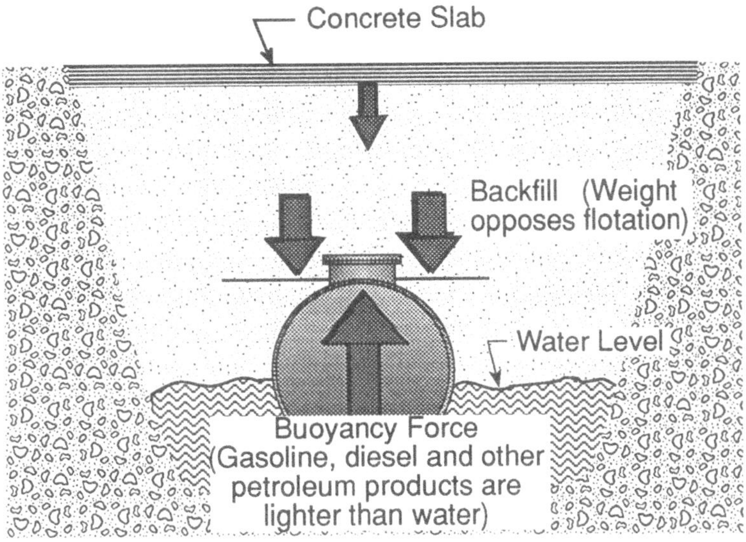

Figure III–6. Depth of Cover in Wet Installations. *Burial depth in wet installations must account for buoyancy and restraining forces acting on the tank. Since tanks are typically buried from three to four feet below grade, supplemental anchorage is usually required only for large diameter tanks.*

☞ Will the excavation still be dry in 10 years? What is the subsurface wet/dry cycle in the area? Can you get an accurate weather forecast for tomorrow?

It is important to check all available records, including historical documents, to note the highest possible water level in the wettest years in order to determine whether the installation has been, is, or will be "wet."

Chapter IV Material Handling

Care in Handling Tanks

Tanks are reasonably durable and are designed for normal handling. However, all tanks must be handled with care to avoid damage to surface coatings, or directly to the tank in the form of punctures, gouges, or scrapes.

☞ **NEVER** wrap chains, cables, or straps around the tank.

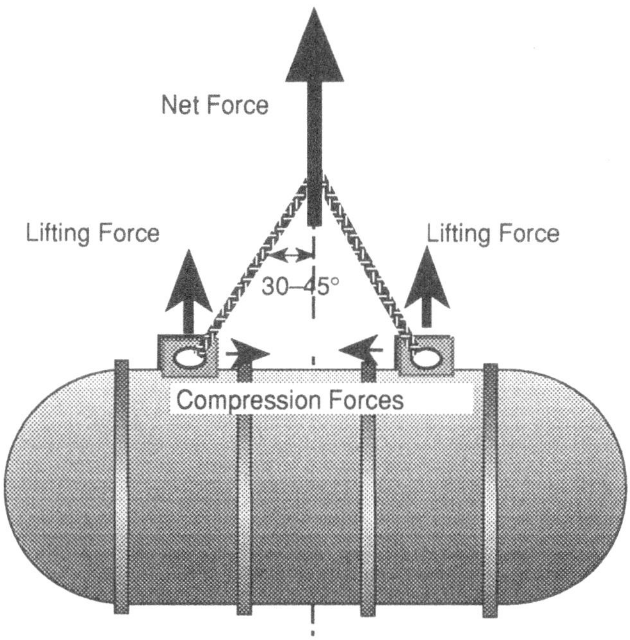

Figure IV–1. Handling Tanks. *Make sure cables or chains used to lift tanks are of adequate length. Cables or chains should be attached to lifting lugs and make not more than about a 30° angle between the cables and a vertical line drawn through the center of the tank. This is to insure that the force on the lugs is an upward force rather than toward the center of the tank.*

Tanks should be moved with care, and must **not** be

- dropped;
- dragged;
- rolled; or
- handled with sharp objects.

☞ Always use guidelines when lifting heavy equipment.

Tanks may be rolled slightly during preinstallation inspection, but installers should be aware that any unnecessary movement can damage coated tanks.

Tanks must be lifted by factory–installed lifting lugs only. Lifting equipment must have adequate capacity; backhoes and forklifts are not adequate to handle most tanks. For small to medium tanks a trackhoe is usually adequate; for large tanks cranes may be required.

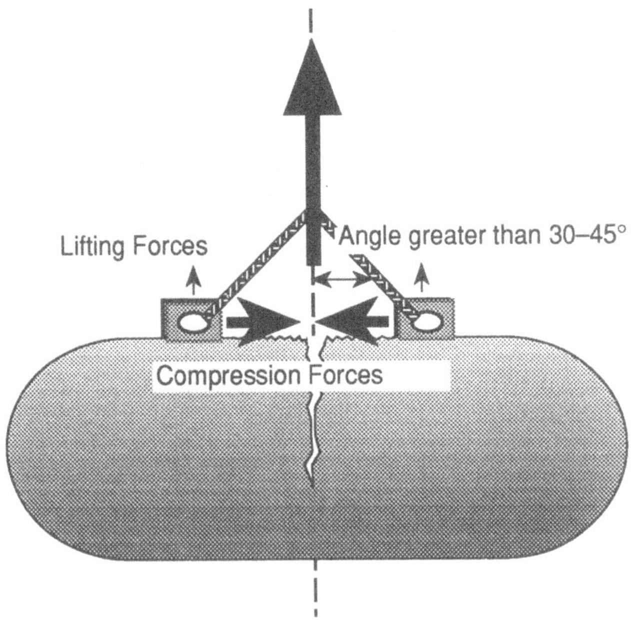

Figure IV–2. Improper Handling Technique. *If the lifting angle exceeds 45°, the lifting forces act on the lifting lugs in a lateral direction, resulting in compression rather than lift. Compressional forces are sufficient to cause structural damage to the tank.*

Cables, chains, or straps used to lift tanks must be of adequate length so that the included angle is no more than about 30° (*See Figure IV–1*). In this configuration the principal forces are lifting forces directed upward to lift the tank, rather than horizontally to compress and possibly damage the tank.

☞ Be especially careful when lifting double–walled tanks.

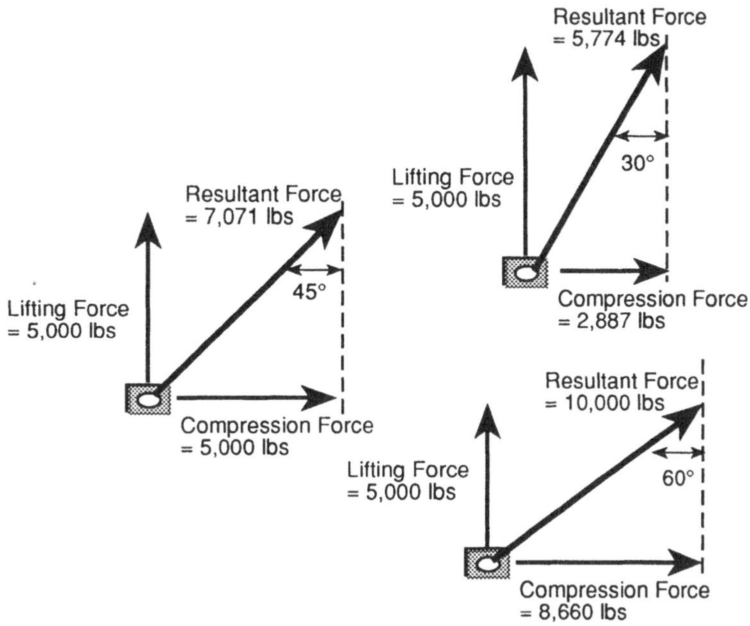

Figure IV–3. Forces Associated with Lifting. *In these examples a vertical force of 5,000 pounds is exerted on each lug of a 10,000 lb tank in order to move it. The net or resultant force is the total force on the cable, and the horizontal, or compression, force is absorbed by the tank itself. The examples show only one lug so the total force on the tank is* ***double*** *the amounts shown.*

If the chain or cable is too short, the angle between the two portions of chain or cable increases, and more of the force of lifting becomes directed in toward the center of the tank, rather than upwards to lift the tank (*Figure IV–2*). The process of lifting then creates compressional forces that tanks are not designed to withstand. In this configuration, it is possible to cause structural damage to the tank — damage which is definitely not covered by the manufacturer's warranty.

> **Example:** *(Refer to Figure IV–3.)* Assume a cable is attached to a 10,000 pound tank with two lifting lugs. In order for the tank to rise, a vertical lifting force that is, at a minimum, equal to or greater than 5,000 pounds must be exerted on each lug. If the cable makes an angle of 45° with the vertical, a simple calculation indicates that an horizontal force of 5,000 pounds is exerted on each lug. If the angle is decreased to 30°, the compression force decreases to 2,887 lbs — the pressure on the tank is reduced by almost half. Going to the opposite extreme, if the angle increases to 60°, the pressure on the tank increases to 8,660 lbs, a 73% increase.

Always attach and monitor guidelines at each end as the tank is lifted. Built–in test wires are **not** suitable for use as guidelines, and factory installed anodes are **not** suitable attachments for guidelines.

☞ Do not use boulders, concrete blocks, cobbles, old engine blocks, or similar objects to chock tanks.

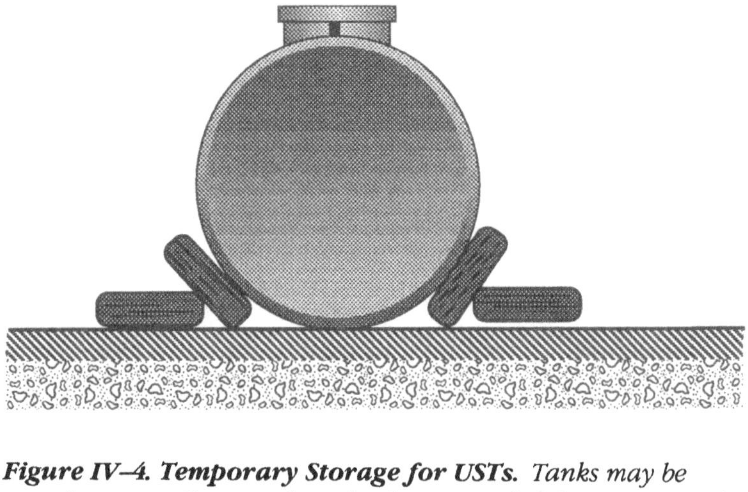

Figure IV–4. Temporary Storage for USTs. *Tanks may be stored temporarily on a clean, level site provided they are properly chocked with old tires or sandbags to prevent rolling.*

Material Storage

Efficient procedures and good workmanship require that the excavation be ready for immediate inspection and installation when the tanks arrive on site.

If for any reason tanks are not to be installed immediately, they may be stored temporarily in a secure area that is free of rocks and foreign objects. Piping and other equipment should be similarly stored in a secure location until assembly and installation. The storage area should be located and constructed so as to minimize the possibility of damage from accidents or vandalism, or injury to workers.

If tanks are to be stored, they must be chocked with tires or sandbags to prevent movement *(Figure IV–4)*. Piping should be stored on wood bats or otherwise protected from contact with the ground. If high winds are expected, tanks should be restrained with ropes of adequate strength. If hail is possible, tanks should be restrained and covered to prevent possible damage.

Preinstallation Inspection

Tanks must be inspected prior to installation. This inspection is the boundary between the manufacturer's or shipper's responsibility and the installer's responsibility. It is very much to the installer's advantage to carry out the preinstallation inspection carefully.

Inspect coated steel and composite tanks carefully. This inspection is important since the coating is an integral part of the tank's corrosion protection, and since small exposed areas of steel corrode very rapidly. The coating should adhere tightly to the underlying structure; even a small gap between coating and structure will eventually corrode. Minor repairs to coatings may be made in accordance with manufacturer's requirements provided the person doing the repairs has been properly trained.

☞ Check to be sure there is good bonding between the coating and steel; i.e., that there are no "bubbles."

☞ Dielectric bushings are often made of nylon. If care is not taken, threaded metal fittings can seriously damage the bushing.

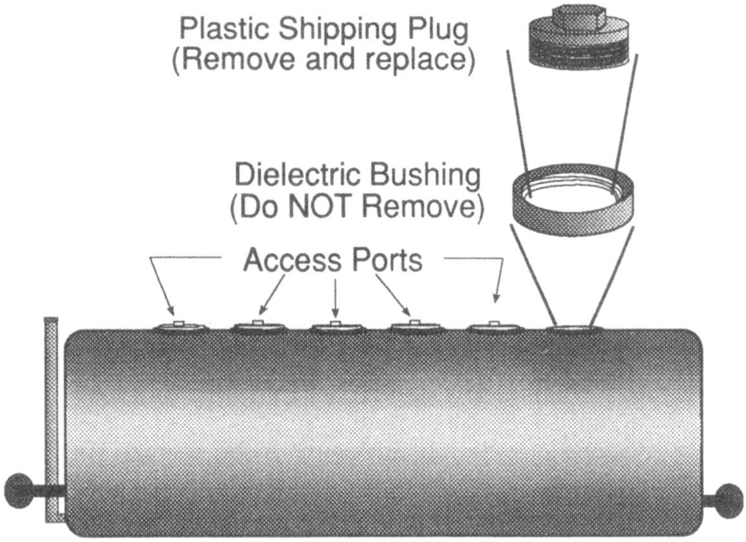

Figure IV–5. Dielectric Bushings. *Steel tanks are shipped from the factory with pre-installed isolation bushings which must not be removed. These bushings are essential to insulate the tank from other installed metallic equipment. The shipping plugs should be removed and replaced with permanent cast iron plugs.*

Preinstallation Tank Testing: Single-Walled Tanks

☞ Always monitor a pressure relief valve to insure it is functioning properly.

1. Remove factory-installed plastic shipping plugs. Do **not** remove dielectric bushings *(Figure IV–5)*. Inspect all bushings for damage.
2. Properly dope and install cast iron plugs in all unused openings. Avoid over-tightening or cross-threading.
3. Install a 10 psig (pounds per square inch, gauge) pressure relief valve in an access port.
4. Pressurize 4– to 10–foot, single-walled tanks to 3 to 5 pounds psig as indicated on an air pressure

gauge which reads 10–15 psig maximum. Hold the tank under pressure for one hour *(Figure IV–6)*.

☞ Air pressure gauges may be in error by 10–15% at each end of the dial.

- Best accuracy for air pressure gauges is at mid-range; therefore, a gauge with a maximum reading of 10–15 psig is strongly recommended for use with an air pressure test.
- Manufacturers recommend a maximum of 3 pounds test pressure for 12–foot FRP tanks and many recommend a maximum of 3 pounds for smaller tanks. Check the specific manufacturer's recommendations before proceeding.
- Install and continuously monitor a pressure relief valve set to 6 psig.
- The pressure inside the tank increases as the temperature increases.

☆ **PRESSURIZED TANKS ARE DANGEROUS. EXERCISE ALL APPROPRIATE PRECAUTIONS.**

Example: A tank is being air tested following a cool night. Early in the morning, when the tank is at 50°F, the tank is pressurized to 5 psig. There are delays and interruptions that delay the soap test until noon. Meanwhile, the temperature increases from 50°F to 100°F. During this time the pressure inside the tank increases to 7 psig, which will exceed the manufacturer's recommendation unless a pressure relief valve is installed and operating.

☆ **AVOID MANWAYS, OPENINGS, AND TANK ENDS WHEN TANK IS UNDER PRESSURE.**

5. Spray a soap solution over the entire surface of the tank, paying particular attention to access opening, manways, welded seams, ribs, or other areas of possible weakness. The tank should be rolled slightly to insure that the bottom of the tank is inspected. Check for bubbles indicating that air is escaping.

☆ **PLACE BARRICADES AND NEVER LEAVE THE TANK UNATTENDED FOR ANY REASON WHILE UNDER PRESSURE.**

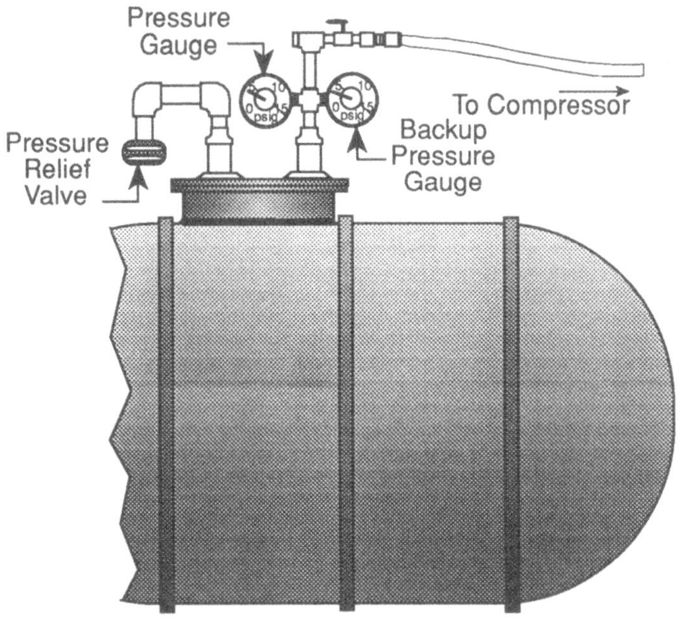

Figure IV–6. Single–Wall Pressure Test. *Single–wall tanks should be air tested at 3–5 psig for one hour. Install and monitor a pressure relief valve to prevent over–pressurizing the tank.*

Preinstallation Tank Testing: Double–Walled Tanks

Since the interstitial space has a relatively small volume that is only about 7% of the volume of the inner tank, it is easy to rupture the outer wall of a double-walled tank if it is pressurized directly from a large compressor. The situation is similar to pressurizing an automobile tire and a bicycle tire; the automobile tire requires a large amount of air to reach 30 psig, whereas the bicycle tire requires only a small fraction of that amount of air to reach 60 psig. A 150–pound compressor that is ideal to pressurize an automobile tire can easily blowout a bicycle tire.

☞ Calibrate pressure gauges regularly to insure accuracy.

Therefore, the procedure for testing double-walled tanks is different from testing single-walled tanks*(Figure IV–7)*.

1. Remove factory-installed plastic shipping plugs. Do not remove dielectric bushings.
2. Install a 10 psig (pounds per square inch, gauge) pressure relief valve in an access port.
3. Properly dope and install cast iron plugs in all unused openings. Avoid over-tightening or cross-threading.

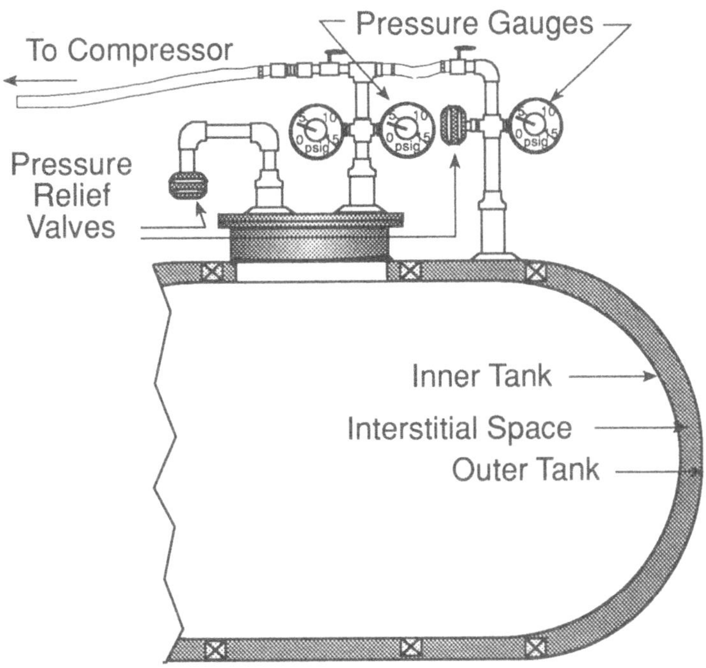

☞ Never pressurize the interstitial space directly from a compressor.

Figure IV–7. Pressure Testing a Double–Walled Tank. *Because the space between the inner and outer wall of a double–walled tank is small, the interstice is pressurized with air bled from the inner tank at approximately 5 psig,* ***not*** *with high pressure air directly from a compressor. Typically the pressure inside the main tank will drop about 0.3 psig.*

4. Depending on the type of tank and the manufacturer's recommendations, pressurize the

inner tank to 3 or 5 psig maximum as indicated on an air pressure gauge with a maximum reading of 10–15 psig.

☞ Do **not** leave for lunch while the tank is pressurized.

5. Close the valve to the inner tank and wait one hour.
6. Pressurize the interstice with air from the inner tank (pressure drops about .3 psig).
7. Close the valve and hold the tank under pressure for one hour.
8. Soap the entire tank and inspect the tank carefully for signs of leaks, paying particular attention to access openings, welds, and joints.
9. Release the pressure.

Other Equipment

☞ Inspecting all equipment before installation can save time and effort.

Equipment and materials should be stored away from the excavation and work areas. Storage areas should be free of debris that could damage equipment, and should be secured against theft or vandalism. Handle all materials carefully to prevent inadvertent damage. Be especially careful with fiberglass pipes and equipment. It is easy to "ding" a 30–foot section of fiberglass pipe if carried carelessly.

➜ See Chapter VII, "**Piping and Other Equipment,**" for pressure testing pipes.

Inspect all piping and fittings before installation, even if it was inspected at delivery. Piping and fittings should be free of dirt, grease, solvents, adhesives, water, and other contaminants. Fiberglass piping and equipment are susceptible to accidental "dings" which show up as whitish blemishes.

Inspect all overfill containment boxes, vents, pumps, vapor recovery equipment, etc. for dents or imperfections before installation. Pay close attention to large pieces of equipment such as fiberglass catchment basins, or long pieces, such as pipe lengths. It is easy to inspect materials before they go into the ground, but, in the event of problems, it is much more difficult to inspect after installation.

Chapter V Backfill

Common Causes of Tank Failures

Prior to the establishment of EPA's technical standards for USTs, it was common practice to use native soil to backfill a tank excavation. Because of this practice several common causes of tank failure were backfill–related. These causes include:

- Using improper, inhomogeneous backfill material;
- Inadequate or improper compaction;
- Rocks or debris left in excavation;
- Voids left under tank;
- Failure to prevent migration of backfill;
- Placing a tank directly on a concrete slab or hard native soil.

☞ USTs are designed to be used *underground* with appropriate backfill support, and must *not* be used as aboveground petroleum storage tanks.

Much of a tank's mechanical support is provided by properly compacted backfill. In general, tanks cannot sustain the weight of overburden or traffic loading without adequate lateral support. The larger the tank, the more critical lateral support becomes. As weight is added to a tank it tends to deflect or to flatten in cross section, unless the backfill provides support. In time this deflection along with flexing of the tank with temperature changes can lead to structural failure and eventual product release.

Proper backfill support is especially important for FRP tanks, which are inherently more flexible than steel tanks. To insure adequate support for FRP tanks, backfill must be properly compacted — either mechanically or by using self–compacting material.

Appropriate backfill material is an important means of corrosion protection for steel USTs since dissimilar soils encourage the formation of corrosion cells. Smooth, small–diameter backfill particles also act to protect epoxy resin and fiberglass coatings that are essential to corrosion protection in steel and composite tanks. Over a period of time as the tanks flex under temperature changes, large diameter material — particularly crushed rock — can abrade coatings and damage FRP tanks.

☞ If the site does not allow room for sloping, consider temporary shoring.

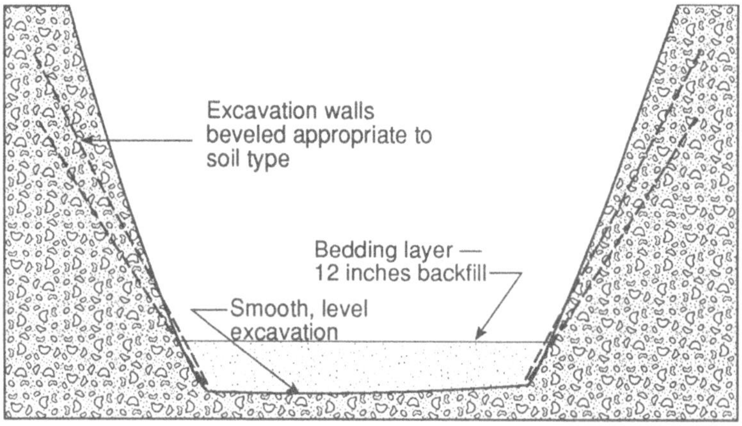

Figure V-1. Basic Excavation. *The excavation should be of adequate size, the bottom should be smooth and level with no debris, and the sides should be sloped according to soil type. Tanks should rest on a 12–inch bedding layer of compacted backfill.*

Placement of Tanks in the Excavation

1. The bottom of excavation should be leveled and carefully inspected for rocks, boulders, or construction debris*(Figure V–1)*. Excavations that are set in native rock should be inspected to be sure that nothing intrudes into the excavation.

2. Tanks should rest on a well-compacted layer of backfill that is a minimum of 12 inches deep. If a hold-down pad is used with a steel tank, the bedding layer may be reduced to 6 inches, if necessary to accommodate equipment requirements.

3. If tanks containing the same product are to be manifolded together, the tank bottoms must be at the same level.

☆ **NEVER ALLOW WORKERS INTO AN EXCAVATION WITHOUT ADEQUATE SHORING OR BEVELING OR BOTH.**

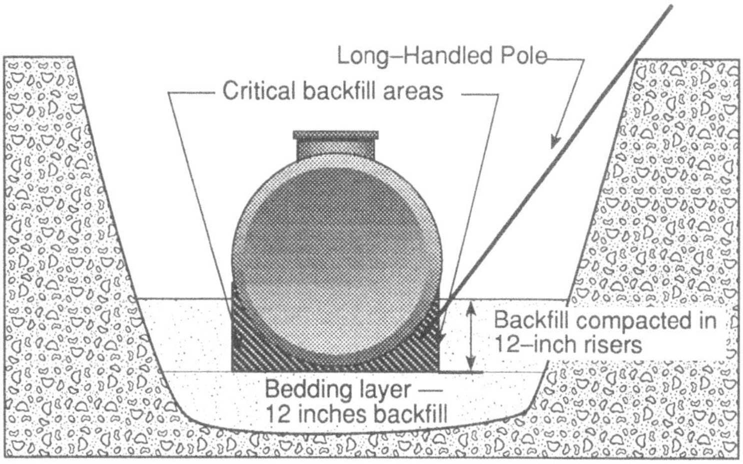

Figure V–2. Critical Backfill Areas. *Backfill under the lower quadrant of the tank must be carefully tamped— either by hand or using a long–handled pole— to eliminate voids in the backfill that could lead to eventual tank failure.*

4. Backfill should be hand–shoveled and tamped around the lower quadrant of the tank to insure that the tank is evenly supported with no voids remaining *(Figure V–2)* to prevent tank slump.

   NOTE: For installations where it may be hazardous for workers to enter the excavation, a long–handled pole should be used to move backfill under the tank to avoid placing a worker in the excavation in violation of OSHA workplace regulations *(Figure V–3)*. For multiple tank installations where the tanks are spaced close together, a long–handled pole may be difficult to use so allow ample room between tanks *(Figure V–4)*.

☞ If it is necessary to add backfill while workers are in the excavation, do so with caution. Workers are not designed to be backfilled!

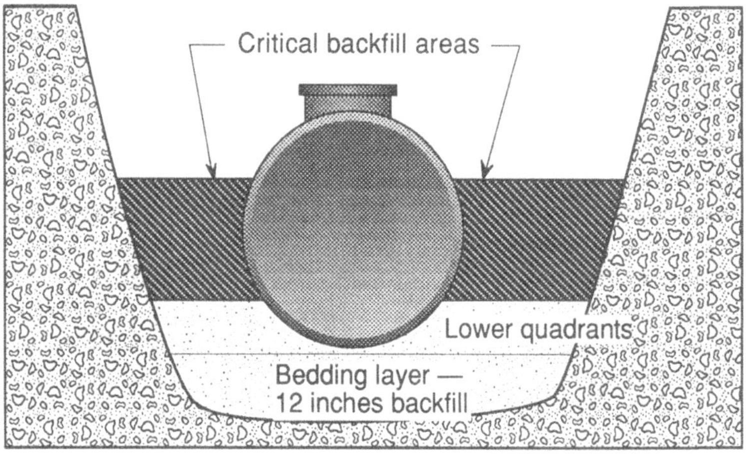

Figure V–3. Critical Backfill Areas. *Lateral support of tanks is crucial to long–term structural stability. If sand is used for backfill, it should be compacted in 12–inch risers.*

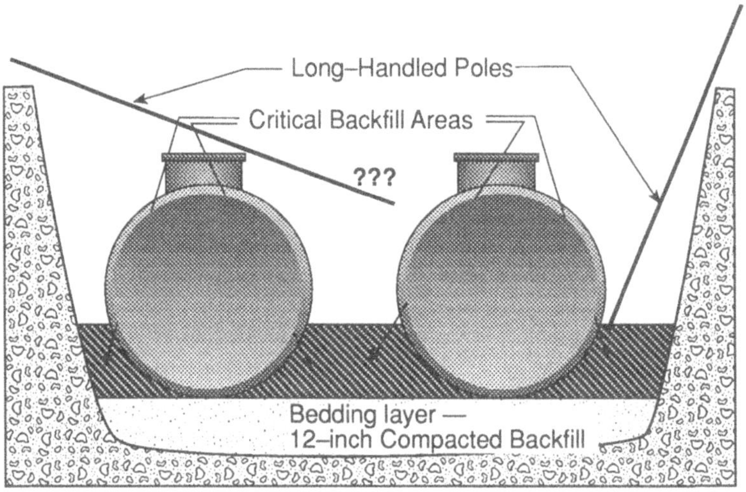

Figure V–4. Critical Backfill Areas in Multi–Tank Installations. *In multi–tank installations care must be taken to insure the elimination of voids under the lower quadrant of the tank. This can generate both safety and construction concerns in installations where tanks are placed close together.*

5. If sand is used as a backfill material, it should be compacted in 12–inch risers to insure the tank is evenly supported, especially around the lower half of the tank.

Ballasting

☆ **BALLASTED TANKS ARE UNSTABLE! NEVER ALLOW WORKERS INTO AN EXCAVATION WITH BALLASTED TANKS UNLESS THE TANKS HAVE BEEN STABILIZED WITH BACKFILL.**

In installations where either of the following conditions exist:

- a high water table in which water will reach any part of the tank;

  or

- surface runoff conditions such that the "bathtub effect" is a factor;

the tanks should be ballasted to prevent flotation before backfilling operations are completed. The preferred method of ballasting is to partially fill the tank with product. Water may be used but then must be removed before completing the installation.

☞ For safety leave the lifting equipment and cables attached until the tank has been ballasted and backfilled.

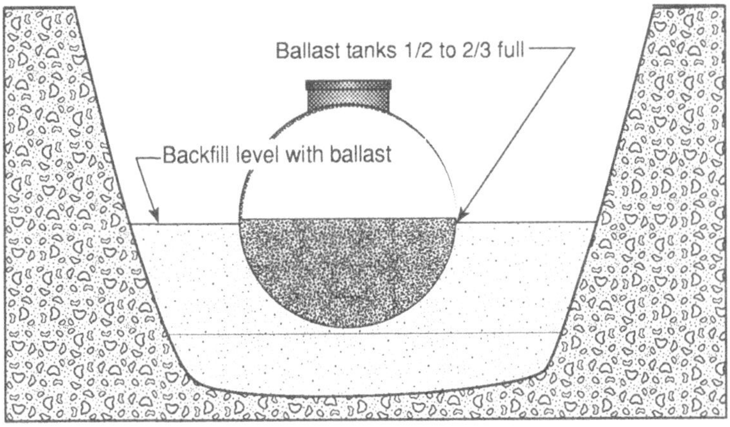

Figure V–5. Ballasting Requirements. *Tanks should be backfilled approximately ½ to ⅔ of the tank diameter, then add ballast to a level even with the backfill.*

☆ **TANKS BALLASTED WITH PRODUCT ARE POTENTIALLY FLAMMABLE AND EXPLOSIVE! TANKS MUST BE PROPERLY VENTED, AND HAVE ACCESS HOLES PLUGGED.**

Ballasted tanks that have not been backfilled are unstable and dangerous. Therefore, backfill must be added to the excavation prior to adding ballast. Product is then added to the tank to a level even with the backfill. Filling tanks to 1/2 to 2/3 full is recommended *(Figure V–5)*.

Once tanks are ballasted with product all necessary safety precautions must be observed, including elimination of all possible sources of ignition.

Backfill Materials: Steel or Composite Tanks

The choice of backfill material depends on a variety of factors including type of native soil, availability of materials, compaction requirements, and cost. Backfill requirements for steel or composite tank installations are:

- Clean, washed, dry, free-flowing, noncorrosive, inert material;
- Particle size should be from 1/8 to 3/4 inch for gravel or 1/8 to 1/2 inch for crushed rock.

☞ If the air temperature is below freezing, it is particularly important that backfill be dry and free–flowing. Frozen backfill can leave voids as it thaws.

River or pea gravel having rounded particles is preferred. Pea gravel is favored among installers since it is self–compacting to about 90% of its compacted density. The disadvantage of pea gravel is that it is not always readily available economically.

Crushed gravel is treated as crushed rock. Neither crushed gravel nor crushed rock compacts as well as pea gravel. If the particle size is too large, crushed gravel can abrade coated surfaces as the tank flexes with temperature changes.

If sand is used, it should be clean and free of debris. Sand should be, but rarely is, compacted in 12–inch risers as it is poured.

Backfill for Nonmetallic (FRP) tanks

☞ Sand is **not** recommended as a backfill with FRP tanks.

Backfill requirements for FRP tanks are similar to those for steel or composite tanks. Suitable backfill materials include:

- Pea gravel: 1/8 to 3/4 inch.
- Crushed rock or gravel: 1/8 to 1/2 inch.
- Not more than 3% backfill material should pass through a #8 sieve.

Sand is no longer recommended by FRP tank manufacturers because sand must be mechanically compacted to provide adequate support for FRP tanks. In the absence of mechanical compaction, tanks are liable to failure due to inadequate support.

Installations in Unstable Soils

1. Installations set in unstable soils, such as sandy soils or unconsolidated alluvial soils, pose a particular problem due to migration of backfill materials into the adjacent soil, and because unstable soils provide inadequate support *(Figure V–6)*.
2. Be sure that excavation sides are sloped sufficiently to eliminate the possibility of slumps or slides. If space is insufficient to permit beveling the excavation sides, adequate shoring must be used.

   Warning: Workers should never be permitted into an excavation in unstable soils without adequate shoring or beveling or both.
3. Provide as much space for backfill as possible — up to 2 feet and never less than 18 inches — between tanks and excavation walls. In the absence of a filter fabric liner unstable or loose

unconsolidated soils can infiltrate backfill and decrease the support available to tanks.

4. A filter fabric should be installed between backfill and unstable soils to prevent soil migration, especially if space requirements do not allow adequate clearance space between tank and walls.

   Warning: Be aware that filter fabric is a thick, heavy material and can hide potential landslides if the excavation walls are not beveled properly. Earth slumps can and have occurred under filter fabric.

☆ **FILTER FABRIC CAN MASK EARLY SIGNS OF CAVE-IN.**

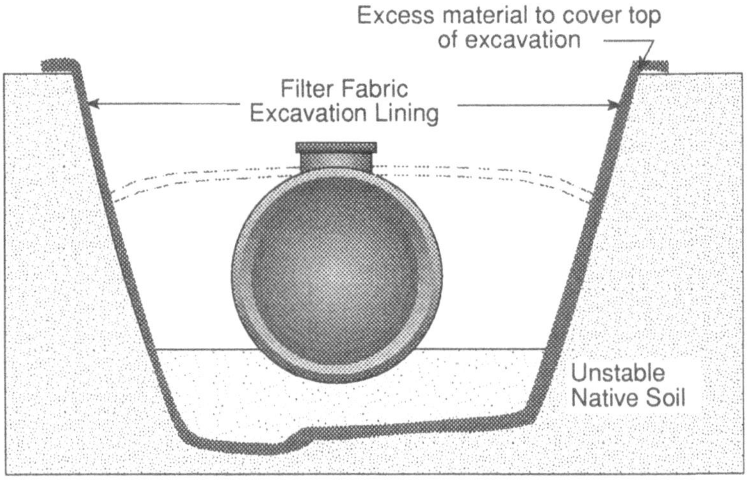

Figure V-6. Filter Fabric. *Use of filter fabric may be necessary in excavations in sandy or otherwise unstable soils. Be sure the excavation is beveled properly to avoid the possibility of the sides slumping under the fabric.*

Chapter VI Supports and Anchoring

Flotation and Anchors

Underground storage tanks are used to store petroleum products. Unfortunately, petroleum products are uniformly lighter than water, which means that USTs tend to float on water *(Figure VI–1)*. Under the right circumstances buoyant forces can be great enough to force the tank up through a finish slab. More often, buoyant forces move the tank just enough over a period of time to damage attached equipment and pipes.

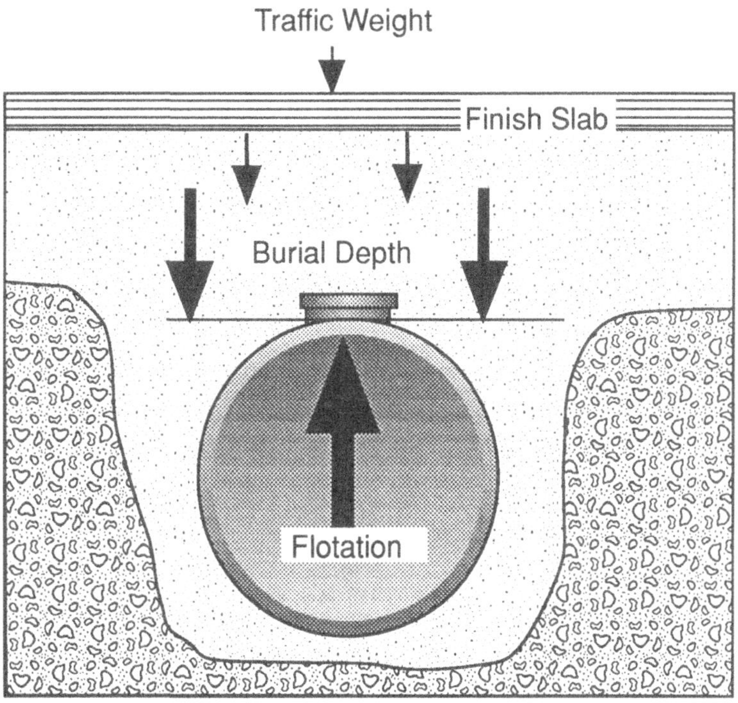

Figure VI–1. Forces Affecting Tank Buoyancy. *All underground petroleum storage tanks float on water because petroleum products are lighter than water. The main restraint against flotation is increased burial depth such that the weight of backfill plus finish slab must be greater than or equal to the maximum buoyancy of the tank at highest possible water level.*

As a result, conditions at many installations will require tanks to be restrained against flotation to avoid damage to attached equipment, finish slabs, or the tanks themselves. Restraint consists of securing the tanks to anchoring devices which effectively increase the total weight of backfill bearing on the UST.

Supplemental anchorage may be necessary under one or more of the following conditions:

☞ Be sure to check historical records to determine the maximum water table possible. Allow for higher levels during rainy seasons or spring runoff.

- In areas of potentially high water table;
- Anywhere high volume surface runoff conditions exist;
- In areas with impervious native soil;
- For large diameter tanks;

  or

- Any other conditions that may lead to tank flotation.

Note that in a wet installation, either with water in the hole, or simply the possibility of water in the hole during the installation, tanks should be ballasted with product as soon as the tank is set in the excavation.

→ Refer to Chapter V, **"Backfill,"** for more information on ballasting tanks.

The main restraint against flotation is to increase the burial depth; that is, to increase the total weight of backfill bearing on the upper portion of the tank. Sand and pea gravel weigh about 60 pounds per cubic foot , reinforced concrete weighs about 87 pounds per cubic foot, and water displaces 62.4 pounds per cubic foot. Therefore, the exact amount of backfill plus finish slab can be calculated for a given size and type of tank. However, the following guidelines are useful indicators:

→ See Reference Section, Recommended Practices for Installation of Underground Storage Tanks, PEI/RP 100-90; Reference 14.

- If the burial depth, measured from the top of the tank to grade level, is at least 60% of the tank diameter, supplemental anchorage is unnecessary.
- However, it is important to note that 60% of a 10–foot diameter tank is 6 feet which is greater than

the normal burial depth in most installations. Therefore, supplemental anchorage is recommended on most installations in which there is any possibility that water, from any source, will reach any portion of the tank.

- Also, be aware that 60% of a 12–foot tank is 7.2 feet, which exceeds the manufacturer's maximum recommended burial depth. Therefore, supplemental anchorage is a requirement on all 12–foot tanks in installation in which water may reach any portion of the tank.

➔ Refer to Chapter III, **"Excavation,"** for information on depth of cover.

Since each additional inch of reinforced concrete equals 2 inches of compacted backfill, additional restraint can be achieved by increasing the thickness of the traffic slab.

Manufacturers recommend that all tanks 12 feet in diameter and larger be anchored if the water table could reach the bottom of the tank under any conceivable conditions.

Types of Anchorage

- Slab at grade and backfill. This is the simplest approach and is effective in areas:
  - with a water table sufficiently deep that water never reaches the excavation,
  - where the excavation is in well–drained native soil, and
  - where surface runoff is not a problem.
- Deadman anchors are typically 12 x 12–inch reinforced concrete beams, which are poured in place to increase the surface area of the tank *(Figure VI–2)*.
- A bottom hold-down pad, which is usually an 8–inch reinforced concrete slab extending 18 inches

beyond the tank sides and 12 inches beyond the tank ends *(Figure VI–3)*.

- Mid–anchoring which consists of a concrete pad poured onto the top portion of the tank.

☞ Deadman anchors must be set outside the tank perimeter. Tanks should never rest on deadman anchors.

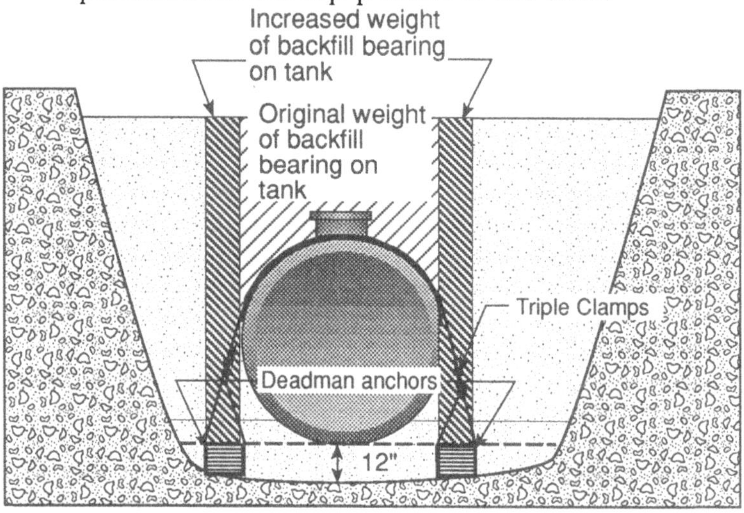

Figure VI–2. Deadman Anchors. *Deadman anchors, in effect, increase the burial depth by increasing the surface area of the tank. The increased weight of backfill resists buoyancy forces. Note the cushion of backfill between tank and anchors.*

The most common types of anchorage, deadman anchors and hold–down pads, work by increasing the effective surface area of the tank. The tank is firmly attached to the anchorage device by hold–down straps; therefore, the total weight of backfill bearing on the tank is increased by the weight of backfill bearing on the area of the anchoring device outside the tank perimeter.

Anchoring devices have the added benefit that the weight bearing *directly* on the tank does not increase. Too much weight bearing directly on the tank surface can cause deflection and structural damage.

☞ A bedding layer of compacted backfill is required between the tank and the hold–down pad or deadman anchors.

Regardless of the means of anchorage, tanks are never set directly on the anchoring device. A minimum 12–inch layer of compacted backfill is laid first, then the tank is set on the bedding layer. If necessary, the bedding layer may be reduced to 6 inches for steel tanks.

☞ A hold–down pad must be larger than the tank perimeter.

☞ Never set a tank directly on a hold–down pad.

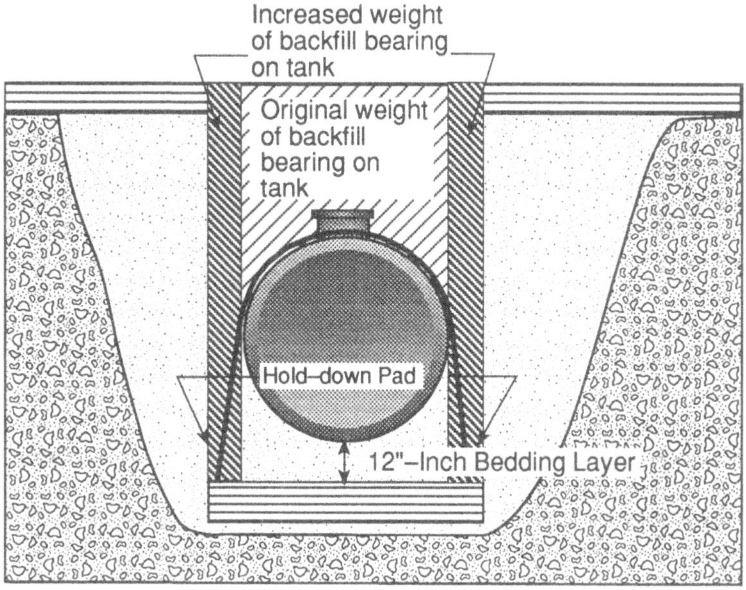

Figure VI-3. Bottom Hold-down Pad. *Hold-down pads are reinforced concrete pads, cast in place, and extending 18 inches past the outside perimeter of the tank. Increased restraint is accomplished by increasing the effective surface area of the tank. Note the 12–inch bedding layer between the tank and the pad.*

Installation of Deadman Anchors and Hold–down Pads

☞ Consult local regulations for hold–down pad specifications.

1. Deadman anchors may be cast separately and placed in the excavation later or cast in place.
2. If more than one deadman anchor is used along a side, beams should butt solidly against each other with no space between.
3. Hold-down pads are cast in place and must be constructed using steel reinforcing rods.
4. Deadman anchors must be set outside the perimeter of the tank; pads must extend at least 18 inches beyond the tank sides and at least 12 inches past the ends of the tanks.

5. Hold–down straps are wrapped around or otherwise secured to anchors and typically clamped with three clamps *(Figures VI–4 and VI–5.*

☞ Consult manufacturers' recommendations for exact number of clamps on hold–down straps.

☞ Tighten straps evenly along the tank.

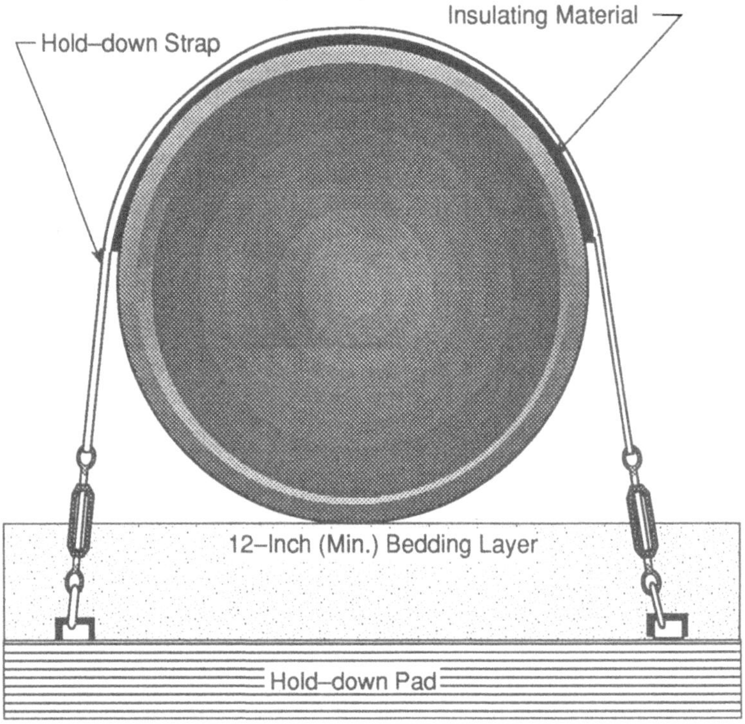

Figure VI–4. Detail of Anchorage: Turnbuckle Assembly. *Typical turnbuckle assembly. Care must be taken to guard against over-tightening hold-down straps and placing excess stress on tanks. Note the protective pads between the hold–down straps and the tank.*

6. In multiple–tank installations each tank must have at least two deadman anchors, one on each side.

7. The tank is attached to the beams or pad by means of appropriate hold–down straps following manufacturer's recommendations.

8. Use caution not to over-tighten straps. Severe structural damage can result from over–tightening hold–down straps. Properly tightened straps must

☆ **OVER–TIGHTENING HOLD–DOWN STRAPS CAN DAMAGE TANKS.**

allow movement and flexing in the tank during normal operations.

9. Tighten straps evenly all along the tank to avoid excess stress on any one portion.

☞ Always use the hold–down kit supplied with the tank being installed. Never substitute hold–down equipment.

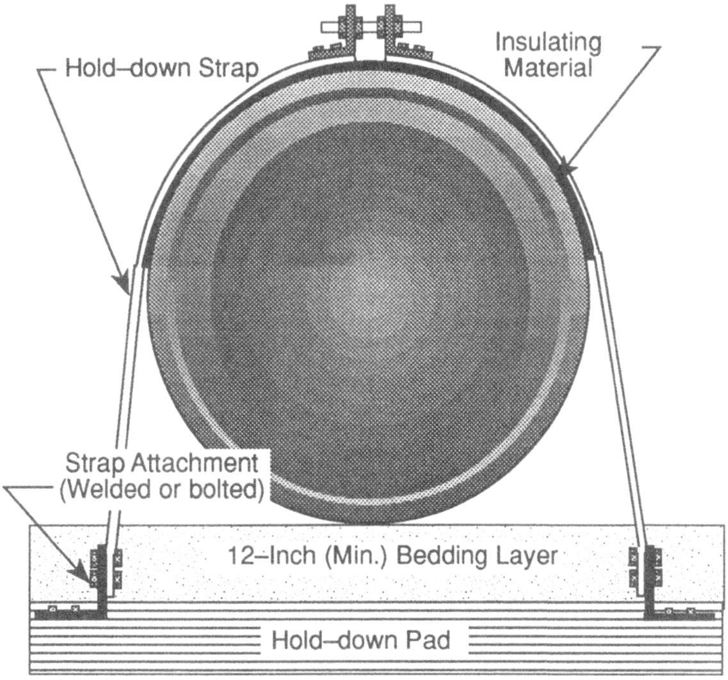

Figure VI–5. Detail of Anchorage: Two–piece Straps. *Alternatively, two-piece hold-down straps may be used. This arrangement is illustrated for steel tanks. Note the use of insulating, protective material between straps and tank.*

10. Follow manufacturers' directions on proper materials and placement of straps, particularly for FRP tanks.

    Note: Some ribbed, single-walled FRP tanks are designed to have straps placed *on* the ribs, some are designed to have straps *between* the ribs, and some are designed to be placed *next to* the ribs. Follow manufacturers' directions closely, and use appropriate hold-down kits.

11. Metallic turnbuckles should be wrapped or coated to protect against corrosion.
12. Straps over steel tanks should be either insulated, wrapped or coated, and cathodically-protected, or be constructed of a non–metallic material.

Installation of hold–down equipment should be approached with care. Ballasted tanks are unstable unless backfilled. Likewise, installations in high water tables or in unstable native soils can present safety hazards to workers in the excavation.

The hold–down strap assembly shown in *Figure VI–5* is particularly advantageous in installations that are potentially hazardous. The turnbuckle arrangement on top of the tank allows workers to tighten hold–down straps without actually entering the excavation itself.

Mid-anchoring Installations

☞ Mid–anchoring is **NOT** recommended by many manufacturers and may violate conditions of the tank warranty.

Mid-anchoring is an alternative to strapping tanks to a pad or deadman anchors. The restraint against flotation consists of a concrete saddle poured across the top portion of the tank. Mid–anchoring was once more common than it is now. Most manufacturers feel that mid–anchoring places too much stress on the tank.

1. Set the tank on a 12–inch bed of compacted backfill.
2. Add backfill around tank until a height equal to ⅝ of the tank diameter is reached, or about 6 feet for a 10–foot diameter tank.
3. Cover the tanks with 2 layers of 30# felt and a layer of polyethylene film.
4. Pour concrete the rest of the way up to the top of the tank, typically 3–4 feet.
5. Backfill the remaining depth to grade.
6. Finish with an 8–inch slab at grade.

7. Manway access ports are recommended with this procedure.
8. Recommended use of this method is limited to single tank installations.

Chapter VII Piping and Other Equipment

The Rest of The System

In an EPA study more than half of all premature UST system failures were due to piping failures and subsequent product releases. It is estimated that, by far, the majority of these failures were due to poor workmanship at the time of installation.

Installation of a piping system can be done in an efficient and durable manner by careful layout and assembly. Time saved by a hurried assembly will almost certainly have to be paid for when the system fails the pressure test or, worse, when a leak develops at a later time.

Piping Layout

An efficient piping system layout minimizes distances and material consumption, enhances operations, and facilitates testing and service. Depth of cover should be deep enough to provide adequate protection from traffic loading, an appropriate slope from dispensers back to tanks, and allow ample room for other equipment. Trench layout should provide ample room between pipes, between pipes and trench walls, and should run no closer than 5 feet from buildings or other structures *(Figure VII–1)*.

Crossovers should be minimized. Where necessary, crossing pipes should not touch to avoid pipe loading. Crossovers are properly made using a tee and 45° elbow.

Product lines should be run in a single trench between tanks and dispenser island. Electrical, vent, and monitor lines should be in another single trench.

Pipe size is determined from pump manufacturer's recommendations for flow rates, number of dispensers, and pump size, but in any case, should not be less than 2 inches nominal inside diameter.

All piping should slope ⅛ inch per linear foot of piping run back to the tank to facilitate draining and testing lines, and to avoid sumps and traps.

☞ Never route product ventilation lines along paths such that the vent lines exit inside a building, under eaves, or into a confined space.

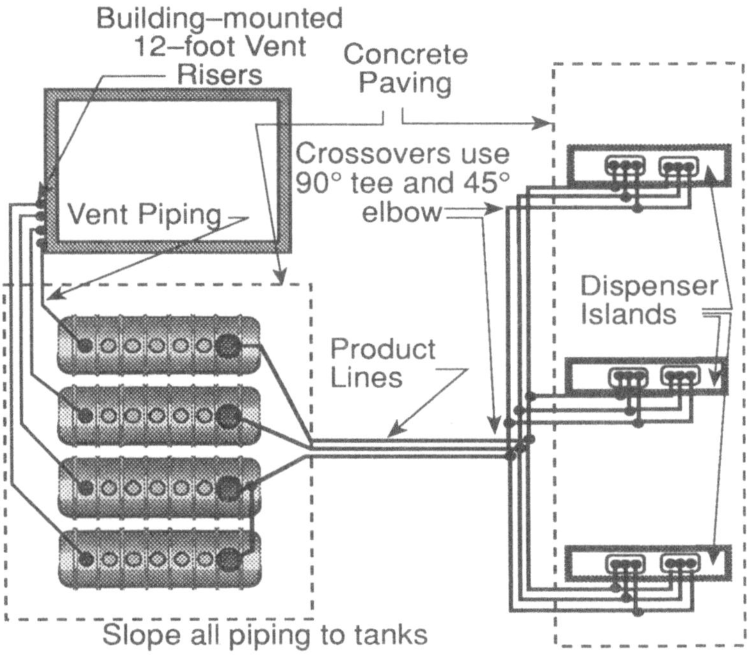

Figure VII–1. Piping Layout. *Typical service station layout minimizes crossovers and overall piping run. Vapor recovery lines should be in their own trench whenever possible.*

Routing should be run:

- in straight lines with as few 45- or 90-degree bends as possible;
- parallel to the tank excavation;

- outside of and parallel to dispensers, and other service facilities.

Materials

Pipes, fittings, and other equipment must be designed to at least equal specifications as the tanks. Piping must be of adequate strength to withstand the pressures of product flow and possible traffic loading. Materials must:

- Be compatible with the product stored.
- Not be damaged by normal operating or test pressures.
- Be resistant to deterioration underground.
- Be isolated from the ground, tanks, and other equipment and cathodically protected, or be constructed of a non–corrodible material.

Trenches and Backfill

Trenches should follow the most efficient routing consistent with the site. Routing should be as straight as possible between tanks and dispensers, and should minimize bends and crossovers.

☞ Product lines should run in a single trench. Electrical lines and ventlines should be routed in a separate trench.

Trench depth should be carefully considered before beginning operations. Allow adequate depth to accommodate piping and any other lines sharing the same trench.

Before pipes are laid, check trenches for rocks and other debris that might damage piping. Pipes should rest on a minimum of 6 inches compacted backfill bedding, which should be smoothed to a uniform grade.

☞ For piping larger than 2 inch i.d. allow 1½ (minimum) to 2 diameters between pipes.

Trenches should be large enough to allow a minimum of one pipe diameter between steel pipe, or two pipe diameters between FRP pipes, with no piping less than

☞ Trenches should be as roomy as possible for the site.

☞ Piping should be buried a *minimum* of 18 inches.

4 inches apart. Installers should take particular care with the following points *(Figure VII–2)*:

- Allow a minimum clearance of 6 inches between pipes and trench walls.
- Piping should be evenly supported on a bedding layer of well–compacted backfill to avoid traps or sumps.
- Avoid traffic loading over unpaved, backfilled trenches. In service station installations it must be assumed that vehicle traffic will eventually occupy all available open space; therefore, all piping trenches should have finish paving in anticipation of traffic loading.

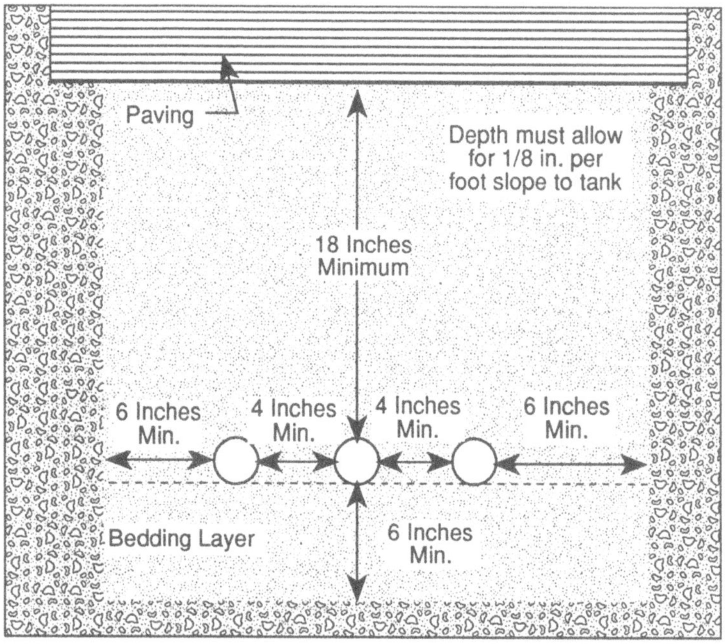

Figure VII–2. Trench Requirements. *Allow adequate space between pipes and between pipes and walls to provide proper support. A minimum of 18 inches clearance between pipes and finish grade is recommended to prevent undue flexing due to traffic weight.*

- A minimum burial depth of 18 inches compacted backfill is recommended. If that is not possible, the following may be used as a guide to *minimum* burial depths:
  - — 4 inches reinforced concrete plus 4 inches backfill,

    or
  - — 2 inches asphalt plus 8 inches backfill.

Installation Procedures: All Piping

☞ Using mechanical compactors over flexible connectors and nonmetallic lines is not recommended.

1. Inspect all materials for signs of damage before proceeding with the installation, particularly where piping has been previously handled with a forklift or similar equipment.
2. Provide adequate clearance between piping and trench walls, and piping and adjacent structures.
3. Avoid crossing lines unless absolutely necessary.
4. Avoid damage during backfilling and compaction.
5. Keep pipe and fittings free of dirt, corrosion, moisture, and debris.
6. Measure and cut pipe lengths carefully. Avoid installing pipe in such a way that it will be under stress.

☞ If methanol–blended fuels are to be used in pipes, always check to be sure that joint adhesives, gaskets, o rings, and all other non–metallic components are compatible. Methanol will dissolve many compounds that resist hydrocarbons.

7. Use compatible pipe dope; wrap ribbon dope with threads to avoid bunching. If the piping system is to transport methanol–gasoline blends be certain that the piping material, joint adhesives, and gaskets are compatible with methanol.

   Note that methanol is an alcohol, and, therefore, chemically very different from petroleum hydrocarbons. Just as methanol will dissolve in water, but gasoline will not, methanol will dissolve materials that gasoline will not harm. The fact that a compound is resistant to gasoline does not mean it is also resistant to methanol. As a general rule, since methanol is chemically more similar to water than gasoline, materials should be inert to both hydrocarbons *and* water.

8. Never remove threaded nylon (dielectric) bushings from openings in steel tanks when installing pipes, fittings, or permanent plugs. These bushings are necessary to isolate a steel tank from the pipe network.

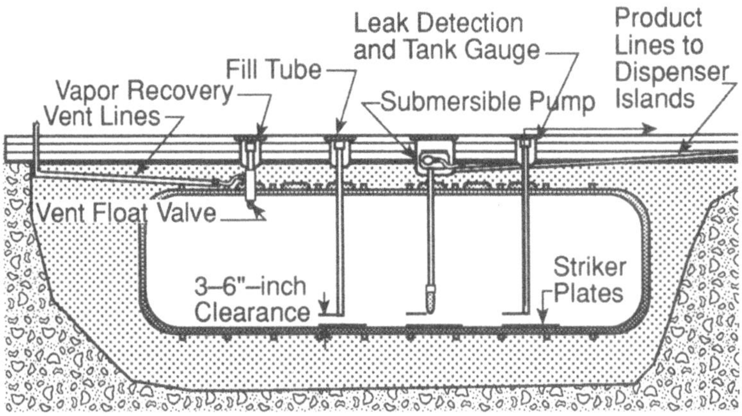

Figure VII–3. Typical Arrangement of Accessories. *Note the use of flexible connectors and catchment basins. An optional arrangement would utilize a catchment basin and a single manway access.*

9. To avoid damage to tank bottoms when installing fillpipes, submersible pumps, or other similar equipment, do not drop the pipe sections into the tank during installation.

   Note: Well–designed tanks will have plates beneath the access ports to protect the bottom from damage.

   - Cut suction lines and fill lines at least 3 to 6 inches from tank bottom — 4 inches is recommended *(Figure VII–3).*
   - Suction stubs should have the same diameter as do the product lines they serve.

☞ Petroleum vapors are heavier than air and will accumulate in low points such as catchment basins and manways. Adequate ventilation is important.

- Use similar appropriate procedures for submersible pumps, monitoring equipment and automatic tank gauge probes being careful to follow manufacturers' recommendations for installation for each specific piece of equipment.

10. Install pumps, check valves, impact valves, flexible connectors, overfill devices, gauges, monitoring systems, and other system components in accordance with manufacturers' recommendations.
11. Submersible pumps must be free to move with the tank. Separate from traffic slab with a street box.
12. Provide adequate ventilation for manways.

Installation Procedures: Metallic Piping

1. The minimum standard for piping should be at least Schedule 40 wrapped/coated black steel pipe.
2. Protect all piping with nonconductive material where it passes through paving or other structures.
3. When cutting threads inspect cutting dies frequently to insure tight metal-to-metal seals. Good seals in threaded steel pipes depends on smooth metal–to–metal contact along the threads. Rough threads will not seat no matter how much dope is used.

☞ Flexible connectors are preferred wherever possible rather than swing joints.

4. Ream and de–burr pipe ends after cutting.
5. Flexible connectors should be used at the beginning and end of each product, vent, and vapor recovery line; and, if necessary, where lines change directions.
6. Installing steel unions underground is not recom-

mended. New designs in composite piping place unions in accessible manifold boxes so that the pipes may be underground, but they are never *in* the ground; i.e.; exposed to backfill and corrosion.

7. Protect metallic piping in contact with the ground with adequate cathodic protection.

☞ Swing joints are heavy and not recommended for use with fiberglass pipe.

Installation Procedures: Nonmetallic Piping

1. If the piping system is to transport methanol–gasoline blends, be certain that the piping material, joint adhesives, and gaskets are compatible with methanol.
2. Swing joints are not recommended for use with FRP pipe; see manufacturer's recommendation for the specific pipe material.
3. Flexible connectors are not required if at least 4 feet of straight run exists between changes in direction.
4. Use care in assembling joints; avoid dirt, grease, and moisture. Wipe joints after cutting to remove dust and chips.
5. Cold weather installation will require extra time or heat for adequate curing. Follow manufacturer's recommendations when using heat packs.
6. Wrap or coat and cathodically protect all metallic components.

Tightness Test For Piping: Procedure

☆ **NEVER LEAVE PRESSURIZED EQUIPMENT UNATTENDED**

Since improper installation is a major cause of product releases from pipes, a careful and accurate pressure test of newly installed piping systems is crucial to a successful installation of an UST system.

☆ **50 PSI CAN BE DANGEROUS! BE SURE FITTINGS ARE TIGHT AND A PRESSURE RELIEF VALVE IS INSTALLED.**

☞ Periodically test pressure relief valves to be sure they are working properly. In operation; always monitor pressure relief valves, especially in very cold weather.

1. Pressure test newly installed piping at 50 psi *before* trenches are backfilled. Follow manufacturers' guidelines for air test procedures.
2. Be sure to install a pressure relief valve and use a pressure gauge with 60–65 psi as the maximum reading.
3. Allow the pressure to build slowly.
4. Make sure the pressure gauge is monitored at all times during the pressure test.
5. Soap joints and fittings. Be sure to work the soap solution around and under joints and fittings.
6. Inspect for signs of leaks, particularly around and under joints and fittings.
7. Hold the pressure for one hour before depressurizing lines. Allow pressure to bleed off slowly.
8. Hydrostatic tests of in–place lines should be at 110% of operating pressure, but not less than 50 psi.

Vapor Recovery and Vent Lines

→ Refer to NFPA 30 § 2–4.5. Reference 25.

All USTs must be vented to prevent the build–up of potentially explosive vapors. Vent lines must terminate a minimum of 12 feet above ground level or two feet above roof line. Normally, vents may not pass through a building *(Figure VII–4).*

Ventline sizes are determined by a combination of flow rates and pipe run.

Many states and localities now require Stage I and/or Stage II vapor recovery equipment installation on all new UST system installations to prevent vapor buildup and to reduce hydrocarbon vapor emission into the atmosphere. If vapor recovery equipment is installed, NFPA–30 requires that ventline not be obstructed by devices provided for vapor recovery of other purposes unless the tank and associated equipment are protected to limit back–pressure buildup to less than the maximum working pressure of the tank.

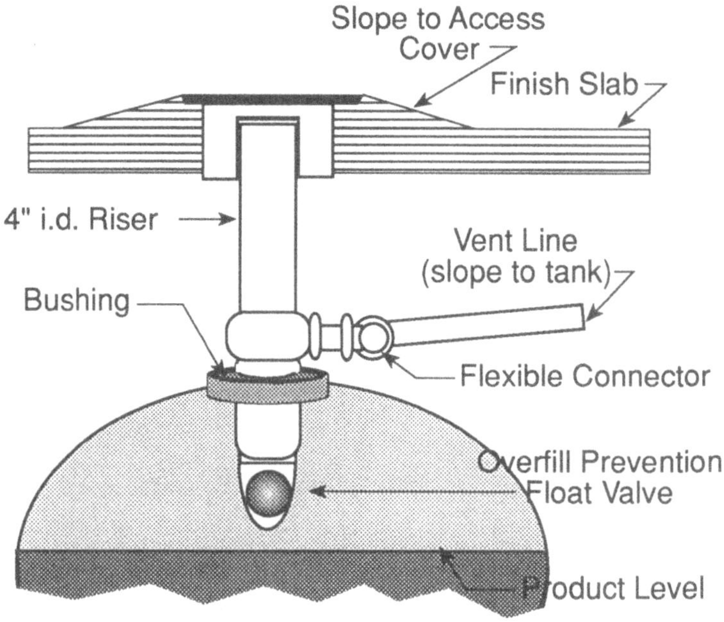

Figure VII–4. Stage I and Stage II Vapor Recovery Installation. *Typical stage I vapor recovery installation with overfill prevention ball check valve is shown. As the tank approaches 95% full, the check valve closes to increase backpressure and shut off fuel flow from the tanker. Stage II lines slope from dispenser islands back to tank. (Tank not to scale).*

Vapor Recovery Equipment

Stage I vapor recovery is designed to prevent release of product vapors into the air at the site of the tank. Product vapors released during tank filling operations are vented back into the tanker. Stage I recovery also acts to condense vapors produced as a result of temperature changes within the tank. Typically, Stage I vapor recovery installations are equipped with overfill prevention check valves *(Figure VII–4 and VII–6).*

Stage II vapor recovery is designed to prevent release of product vapors at the site of the dispenser. It appears that Stage II vapor recovery devices will be required by the Clean Air Act of 1990 on dispensing operations in non–attainment areas .

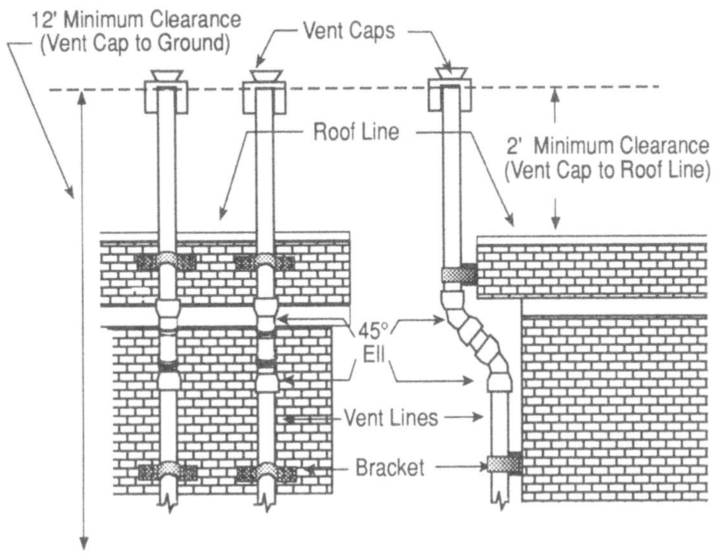

Figure VII–5. Building Mounted Vents. *Vent lines must conform to local building codes; the arrangement shown here is typical. Note that vents should terminate a minimum of 2 feet above adjacent roof lines.*

➔ See also NFPA 30, Chapter 3 for general piping and vent line requirements.

Vapor Recovery and Vent Lines: Installation

1. Vent lines must be of adequate capacity; 2" vent piping is recommended. Piping should not be less than 1¼ inch i.d. *(Figure X–4, p. 112).*
2. Vent lines must be located so that the discharge point is outside buildings, not under an overhang,

higher than the fillpipe opening, and not less than 12 feet above ground level or 2 feet above the roof line if attached to a building *(Figure VII–5)*.

3. Vent outlets must be protected from blockage, and not obstructed by other equipment.

4. Vent piping for tanks storing Class I liquids may **not** be manifolded with tanks containing Class II or III liquids.

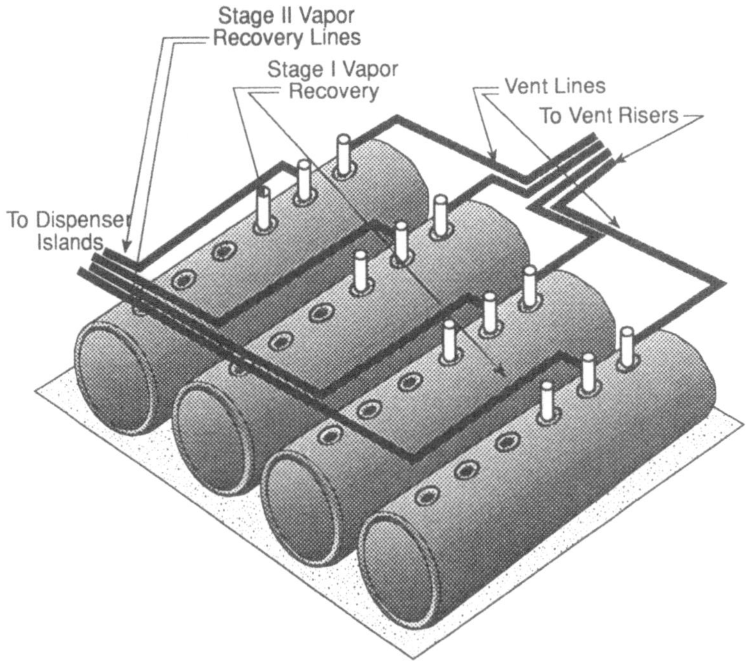

Figure VII–6. Stage I and Stage II Vapor Recovery Installation. *A typical Stage I and Stage II vapor recovery installation. Stage I vapor pickup uses one of the access ports which should be convenient to the fillpipe. Stage II lines run to dispensing islands and should slope back to tanks. Stage II vent lines must not be manifolded with tanks containing different products. All vent lines should run in their own trench and be protected from traffic loading. Vent assemblies should be fitted with check balls as shown in Figure VII–4.*

5. Vent lines should be a minimum of 12 inches below finish grade and protected from traffic loading.
6. Vent pipes should slope ⅛ inch per foot back to the tank. Where vent lines pass through concrete or asphalt, the lines should be protected.
7. Aboveground piping should be Schedule 40 galvanized steel and protected from possible damage by traffic.

Manways and Equipment Access Ports

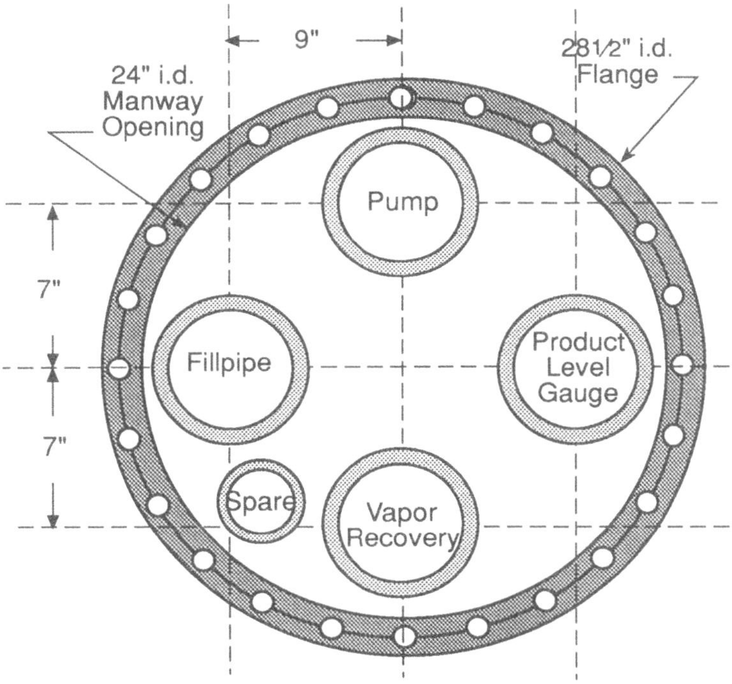

Figure VII–7. Typical Equipment Access Opening Layout. *The arrangement shown is typical for either FRP, steel, or composite designs. Equipment access ports are convenient for service and maintenance. All such access should have adequate ventilation.*

Increasingly, new tanks are fitted with manway access ports to facilitate service and maintenance (*Figure VII–7*). Manways for tank access are similar and may be equipped with ladders for convenience. A single large capacity catchment basin is used for spill prevention.

Chapter VIII Tank Removal and Closure

Safety

There is no question that tank removal is the most hazardous aspect of UST work. The **Safety Letter,** published by Petroleum Equipment Institute, PEI, has documented story after story of workers injured or killed as a result of fire and explosions of used petroleum storage tanks.

All such work must be accomplished in accordance with federal, state, and local requirements as well as accepted safety standards and plain common sense.

☞ A full tank may be less explosive than one that is nearly empty, but it is equally flammable.

It is absolutely essential to remember that USTs are used to store petroleum fuels that are highly flammable or explosive, and to treat all storage tanks with the utmost caution. As with firearms, the "empty" tank is often the most dangerous. Since only vapors, not liquids, will burn or explode when mixed with oxygen, tanks that are completely full of product are much less explosive than a tank that has been *almost* completely emptied.

All petroleum storage tanks accumulate sludge — a heavy, usually rather viscous, residue of oxidized and polymeric hydrocarbons — in the tank bottom. Sludge contains dissolved petroleum products which readily outgas to form vapors when the tank is moved, heated, or just left alone for a period of time. Similarly, petroleum products become trapped in seams and joints, and will outgas over time. Tank reliners consider a tank "safe" only after the interior has been sandblasted down to bare metal.

The following incident appeared in **Safety Letter,** May, 1990:

> "A self–employed scrap iron and metal dealer was killed while 'scraping' an abandoned underground

storage tank. The man was working alone and using an acetylene torch to cut the tank into scrap when the flame from the torch ignited fumes inside the tank and touched off an explosion. The force of the blast lifted the 10,000–gallon tank into the air, sending it about 50 feet from its initial spot. A tank end was blown about 450 feet into a nearby field...

The victim's brother said the worker was experienced in cutting scrap metal and 'knew better than to cut up a gas tank.'"

Petroleum Toxicity

☆ **PETROLEUM VAPORS PRESENT BOTH EXPLOSIVE AND TOXICITY HAZARDS TO HUMAN HEALTH.**

Not only are petroleum fuels flammable and possibly explosive, the vapors are also hazardous to human health. Therefore, the tank remover must take all possible precautions to avoid prolonged worker exposure to petroleum hydrocarbon vapors. Toxic effects may be acute — the symptoms will show up within a few minutes to a few days, or chronic — the symptoms may not show up for weeks or months. Since petroleum vapors are heavier than air, they will tend to accumulate in the lowest point of the excavation, especially on a hot, still day. Supervisors should be alert for the following symptoms:

- Acute hydrocarbon toxicity may result in symptoms of intoxication, including dizziness, loss of coordination, and unconsciousness.
- Should a worker show signs of dizziness or loss of coordination, immediately remove the individual to fresh air. Obtain medical attention as soon as possible.

Remember that vapor pressures of hydrocarbon solutions increase with temperature. Vapors are more of a problem on hot days than cold days.

Work crews that do removals on a regular basis need to be aware that long–term exposure to hydrocarbon vapors can be harmful.

- Several components of petroleum fuels contain compounds, such as benzene and xylenes, that can result in chronic, long–term effects. Worker exposures should be carefully monitored to remain within acceptable exposure limits.

A combustible gas indicator, operated by trained personnel, should be used to monitor concentrations of potentially explosive vapors. The tank atmosphere should be at less than 20% of the lower explosive limit (LEL =1.4% gasoline vapors by volume) before work proceeds.

CAUTION: Even though residual product has been removed, tanks cannot be guaranteed to remain vapor-free, and must be treated appropriately.

Removal Procedures

☞ Before removal, notify the appropriate local authorities — often the Fire Marshal or local fire district.

1. Drain piping back into tank. Disconnect piping from tank. Cap or remove piping.

   CAUTION: All sources of ignition must be eliminated. This includes all smoking materials, electrical equipment, and internal combustion engines.

2. All remaining product *and residues* must be removed before any work is done on the tank.
   - Use explosion-proof or air-driven pumps.
   - Pump motors and suction hoses must be bonded to the tank or grounded to prevent spark-induced explosions.

☞ *Purging* means removing all flammable vapors from the tank; *inerting* means removing the oxygen from the tank.

- The area around the truck must be kept vapor-free by using proper ventilation procedures.

CAUTION: All equipment must be grounded or firmly bonded to the tank to prevent electrostatic buildup (sparks) and possible explosion.

3. Begin excavation and expose the top of the tank.
4. Remove ancillary equipment, except a single vent line, which should remain connected until inerting is completed.
   - Alternatively, install a threaded 12–foot galvanized pipe in the access port opposite the one to be used for the inerting process.
   - In either case, the vent should terminate at least 12 feet above the top of the tank.
5. Temporarily plug remaining tank openings.
6. The tank must be inerted or purged of all flammable vapors by one of the following methods:
   - Adding 1½ pounds of dry ice (solid CO_2) per 100 gallons of tank capacity.

☆ **NEVER HANDLE DRY ICE WITH BARE HANDS.**

Warning: Handle dry ice with care as it can cause severe skin burns.

   - Purging the vapors by means of compressed air or fan ventilation.
   - By passing an inert gas such as N_2 or CO_2 through the tank under low pressure.

☆ **NEITHER N_2 NOR CO_2 WILL SUPPORT LIFE.**

7. In the event workers must enter the excavation remember that petroleum hydrocarbon vapors are heavier than air and will accumulate in the lowest portion of the hole.

CAUTION: Workers should enter the excavation during purging operations only if absolutely necessary, and then with strict observation of appropriate safety procedures.

8. Inerting or purging must be done in the time period immediately prior to removal.
   - Adequate ventilation must be maintained during purging operations. Vapors should be vented at least 12 feet above grade and 3 feet from adjacent roof lines.
9. The tank should be tested at periodic intervals for explosive limits up to the point that the tank leaves the site.
10. Depending on local safety regulations, one or more openings may be cut in the tank as needed.

    Warning: If it is necessary to send a worker into a tank, and dry ice or an inert gas has been used to purge the tank, the interior will be **oxygen deficient**. External or self–contained breathing equipment and a safety line are absolutely mandatory. Support personnel must monitor the safety line at all times.

11. Remove the tank from the excavation using lifting equipment of adequate capacity. The tank may be cut into sections as needed to expedite removal. Check local regulations regarding the necessity of cleaning tanks prior to cutting.

This incident was reported in **Safety Letter**, October, 1989:

> "One of four 10,000–gallon underground steel storage tanks exploded as workmen were preparing it for removal.
>
> Two workmen were unbolting the four tanks' manhole covers with an acetylene torch in order to facilitate access to their interiors. The bolts on two of the tanks had already been cut, and the men were cutting away the studs of the third tank when

the tank exploded. The 35–year–old worker who was performing the hot work on top of the tank was hurled about 20 feet into the air and out of the excavated pit surrounding the exposed tank top. He suffered third–degree burns over much of his body.

The second workman, who was walking to a pickup truck at the time of the explosion, received moderate burns to his upper body. However, he managed to put out the lingering vapor fire that followed the blast with a portable extinguisher before emergency services personnel arrived.

☞ Tanks that have been emptied of product but not cleaned to bare metal may degas enough vapors to cause an explosion.

The fire chief interviewed one of the workmen later that day regarding the sequence of events that led up to the accident. Apparently, the workmen never purged the tank of flammable vapors or monitored the surrounding area with a combustible gas indicator before or during the hot work."

A final quote from **Safety Letter**, January, 1990:

"The business of removing and disposing of underground petroleum storage tanks is *dangerous* when proper procedures are not followed!" (Emphasis added).

Disposal

Used tanks must be disposed of in accordance with applicable state and local regulations. Some areas allow tanks to be reused or recycled. In areas that do not allow used tanks to be reused, a tank should be perforated to render it unsuitable for future use.

☞ When transporting a used UST from a site, there is a potential fire or explosive hazard; plan routes accordingly.

If the tank has *not* been cleaned to *bare metal* before leaving the site, that is, if there is residual petroleum product remaining in the tank, the tank is a potential bomb and should be handled with great care. Route selection for transporting the tanks should be chosen so as to minimize travel through residential areas.

Tanks and transportation equipment must meet and follow applicable DOT and state regulations. The used UST should be labeled appropriately according to local regulations. The following labeling, in 2–inch high letters, is typical:

TANK HAS CONTAINED { ***TYPE OF PRODUCT*** **}**

NOT VAPOR FREE

NOT SUITABLE FOR STORAGE OF

FOOD OR LIQUIDS INTENDED FOR

HUMAN OR ANIMAL CONSUMPTION

DATE OF REMOVAL: {*MONTH/DAY/YEAR*}

☞ Regulations for tank labeling vary from one jurisdiction to another. Be sure to check local regulations.

If the tank has contained, or is suspected of having contained leaded gasoline in the past, an additional label should be added as follows:

TANK HAS CONTAINED LEADED GASOLINE

LEAD VAPORS MAY BE RELEASED IF HEAT IS APPLIED TO TANK

In many jurisdictions an inspector must visually check the tank for signs of leaks before it is removed from the site.

A bill of sale should be used to indicate transfer of ownership of the used UST. The contractor should also provide the owner/operator with proof of proper disposal. Since the previous tank owner's potential liability does *not* end with removal, it is in the owner's interest to know what became of the tank.

Frequently the final resting place of a used UST is a scrap yard where it may sit for years. Such areas should be treated as bomb disposal sites, and rigorously protected from public access. A tank that has been drained of product, but not sludge, or drained of sludge, but not cleaned, is simply a bomb waiting to go off.

Site Assessment and Closure

☞ Regulations for tank closure vary from one jurisdiction to another. Be sure to check local regulations.

Before the tank installation is closed, the site must be demonstrated to be free of contamination. The surrounding soil must be tested for the presence of petroleum products in accordance with applicable state and local regulations.

- Test samples should be taken from several locations under the tank location in native soil, not just the old backfill. Particular attention should be paid to areas immediately under fill pipes.
- Monitoring wells should be set into groundwater downstream from the tank location.

☞ Soil sampling must be done carefully with due regard to cost, statistical validity, chain of custody, and documentation.

- Soil samples should be taken with regard to proper sampling technique and chain of custody. This topic, which unfortunately is beyond the scope of this book, is extremely important, not only to eventual remediation choices, but also in possible litigation.
- Copies of laboratory analysis and the site assessment must be submitted to the appropriate implementing agency.
- If there is evidence of product release into the environment, the owner/operator must report the release to the implementing agency and take appropriate actions to remediate the contamination.

Tank Closure in Place

Tanks that are no longer useful should be removed and disposed of in a proper manner. However, in certain unusual circumstances it may be physically impossible to remove a tank without unacceptable damage to surrounding structures, as, for example, the tank that has become embedded in a building foundation. In such cases it may be acceptable to close the tank in place using the following procedures:

☞ Closure in place does not negate the requirement for a site assessment, including sampling directly under the tank. Check local regulations.

1. Drain and disconnect piping. Cap or remove the pipe network.
2. Disconnect and remove all attached equipment from the tank.
3. Remove all product from the tank following appropriate safety procedures.

   CAUTION: Properly ground all equipment during purging operations to prevent static charge buildup.

4. Purge the tank of remaining flammable vapors, taking all appropriate precautions to prevent static buildup and a possible explosion.
5. Depending on local regulations, one or more adequate openings are cut in the top of the tank. If it is necessary to place a worker inside the tank to complete cleaning operations, be sure to follow all applicable safety procedures, including safety line and breathing apparatus.

   CAUTION: If it is necessary to put a worker inside the tank to complete cleaning operations, follow all appropriate safety precautions

6. Clean remaining sludges from the tank bottom.
7. Backfill the tank completely with a sand-slurry or a cement–slurry mixture.

8. Perform an adequate and complete site assessment to satisfy the requirements of the local regulatory agency, including soil samples from directly beneath the tank.

9. This procedure is not recommended as a general practice. It is available when actual removal is not an option.

Chapter IX Corrosion Protection

Design Criteria For Buried Steel Structures

It is widely recognized that corrosion is a major cause of releases from existing USTs. In 1989, approximately 83% of existing USTs were unprotected (bare) steel tanks, at least half of which had no form of corrosion protection at all. Data from EPA surveys indicate that between 60 and 70% of all leaks from existing steel underground tanks are caused by corrosion. Thus, *Technical Standards* requires that all new installations have corrosion protection installed and maintained on a regular basis.

Standard specifications for coated and protected steel tank construction include:

- *Specification for STI–P3 System of External Corrosion Protection of Underground Storage Tanks,* STI–P3–87;
- *Corrosion Protection for Underground Storage Tanks,* UL- 58;
- *Cathodic Protection of Underground Petroleum Storage Tanks and Piping Systems,* API–1632.

The EPA's findings indicate that the piping portion of UST systems is at least as likely to fail as the tanks themselves, particularly where the failure is premature (leaks occurring in newer installations). Piping failures are almost uniformly caused by either poor workmanship or corrosion. *Technical Standards* requires that all steel piping in contact with the soil and carrying product to or from tanks must have cathodic protection. Metallic components that are not in contact with the soil do not need to be protected. Indeed, new commercial double–walled piping systems take advantage of this by placing all metallic components underground but not in the ground.

➔ See Appendix A, **"Batteries and The Corrosion Reactions of Iron,"** for more detail on corrosion cells.

Corrosion in Buried Steel Structures

Corrosion is the tendency of a metal to return to its natural state through a spontaneous electrochemical reaction with its environment. Iron is produced from naturally occurring oxide ores such as hematite, Fe_2O_3. Thus, when iron corrodes it becomes the anode in a corrosion cell and is oxidized back to "rust", which has the approximate formula, Fe_2O_3.

☞ Metallic fittings must be electrically isolated from pipes, tanks, and each other.

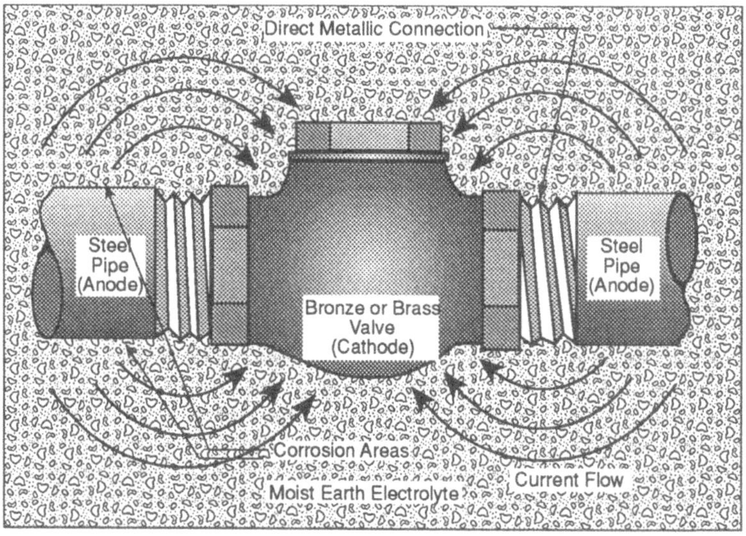

Figure IX–1. Corrosion Due to Dissimilar Metals. *Corrosion cells occur whenever dissimilar metals are in direct contact and a return current path exists. In this figure corrosion cells are set up between steel pipes and a bronze fitting since the fitting is not electrically isolated from the pipe. Differences between steel pipe and welds or solder materials can also set up corrosion cells.*

In buried steel structures corrosion occurs through the formation of **corrosion cells**, which are exactly analogous to batteries. The four essential parts of a corrosion cell are:

- an anode, where corrosion occurs;
- a cathode, where reduction occurs;
- a direct metallic connection to provide a path for current flow;
- an electrolyte solution to provide a path for return current flow that completes the electrical circuit.

☞ Steel contains carbon (a reducing agent) as a hardening additive. Also, steel may contain small percentages of various alloy metals such as vanadium, chromium, molybdenum, nickel, cobalt, and other metals to achieve special properties.

Steel is inherently inhomogeneous since it is composed of about 90% iron and the remainder a mixture of carbon and other alloy metals, such as chromium, vanadium, and nickel. Thus, three components of a corrosion cell are present in steel on a microscopic scale: two dissimilar metals (an anode and cathode), and a direct metal contact. The only component missing is an electrolyte solution, which is provided by a water drop or moist soil.

In UST systems corrosion cells form under one or more of the following circumstances:

- Two different metals in contact *(Figure IX–1)*;
- Variations in the chemical properties of the backfill (oxygen content, groundwater composition, etc.) *(Figure IX–2)*;
- Variations in the physical properties of the backfill (dissimilar backfill materials);
- Stray electrical currents from adjacent steel structures *(Figure IX–3)*.

The first condition is inherent in steel. Therefore, one of the principal defenses against corrosion in UST systems is to apply a uniform, carefully bonded coating to tanks and piping. In the STI– P3 design the coating is a ⅜–inch thick coating of epoxy resin; in composite tanks the coating is a one–inch thick layer of fiberglass resin.

☞ Current flow has no inherent direction. It is customary to speak of current flowing from anode to cathode, however, this is only a convention, and has no intrinsic significance.

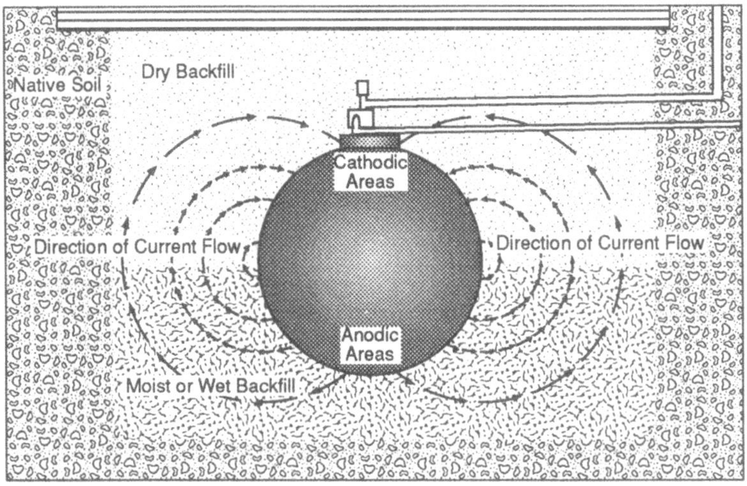

Figure IX–2. Corrosion Cell in Dissimilar Soils. *Differences in physical or chemical composition of backfill materials will induce current flow. Differences may be in water or oxygen content or inhomogeneous backfill.*

☞ Stray currents are generated by any electromechanical or electromagnetic operations; however, the most common and potentially the most damaging are currents generated by grounded DC electric power operations.

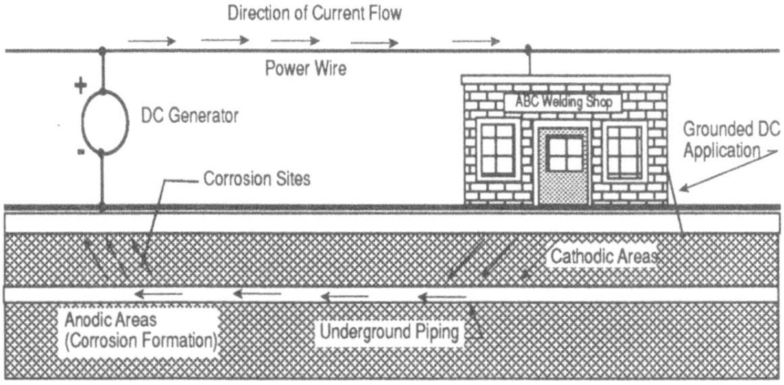

Figure IX–3. Stray Current Corrosion. *Grounded electrical equipment induces electrical currents which follow low resistance paths through the ground and buried steel structures such as pipes and tanks. Corrosion occurs at anodic areas where the current leaves the surface of the structure.*

The second and third conditions (variations in the physical and chemical properties of the backfill) can be reduced by using clean, electrically neutral backfill as described in Chapter V, and by isolating the steel components from the environment, either by coating or double–walled construction.

Since corrosion in buried steel structures is typically characterized by large cathodic areas and relatively small anodic areas, it is important to check the coating of tanks and pipes carefully before installation; a small defect will act as an anodic area to concentrate the soil current and greatly accelerate the rate of corrosion.

Sacrificial Anode/Cathodic Protection

Since corrosion occurs only at the anode and never at the cathode, the most common form of protection for underground steel structures is *sacrificial anode cathodic protection*, sometimes called *galvanic anode* protection.

A metal may be protected from corrosion by forcing it to be the cathode in a planned corrosion cell. This is accomplished by placing it in contact with a metal more negative than it is; the more negative metal becomes the anode and corrodes preferentially.

Both magnesium and zinc are more negative than iron and therefore oxidize (corrode) more rapidly. A *planned* corrosion cell is set up in which magnesium (or zinc) becomes the anode and the tank, or other buried metallic structure, becomes the cathode. If there is a direct metallic connection between the anode and cathode, the anode corrodes preferentially to protect the buried structure.

☞ Anodic protection is simply a planned corrosion cell in which the anode is sacrificed to protect the underground structure.

This type of planned corrosion cell is called **sacrificial anode/cathodic protection.**

In the STI–P3 design prepackaged, attached anodes are magnesium alloy or zinc welded to the tank. The wrapping contains an electrolyte material composed of 75% gypsum, 20% bentonite, and 5% sodium sulfate.

The choice of anode type depends on soil resistivity; magnesium anodes provide higher current and stronger driving force than zinc, but also have a shorter life. Typical lifetimes range from 10+ years in corrosive soils to 50+ years in slightly corrosive soils.

Advantages and Disadvantages

Advantages:

- No external power supply is required.
- Installation is simple.
- Low cost and maintenance.
- Interference from stray currents is rare.
- Efficient for small electrically isolated coated structures.

☞ Inspect coated steel tanks carefully for "holidays" since small areas of exposed steel will corrode rapidly and shorten anode life.

Disadvantages:

- The range of sacrificial anode/cathodic protection is limited since the iron/magnesium (or zinc) driving potential is low.
- May not be practical in soil of very high or very low resistivity.
- Not applicable for large, bare steel structures.
- Anode life may be short. In corrosive soils (200 ohm cm resistivity) magnesium anode life may be as short as 10 years.

Installation of Cathodic Protection Systems

☆ **WELDING TANKS THAT CONTAIN PRODUCT ARE EXTREMELY HAZARDOUS.**

1. During handling and installation of tanks and piping it is important to use care not to damage protective coatings on built–in anodes.

   CAUTION: Use all appropriate precautions when attaching anodes to tanks ballasted with product.

2. Inspect coated tanks for damage and repair all holidays according to manufacturer's recommendations since small exposed portions of a steel surface corrode very rapidly.

3. Be sure to remove the plastic wrapping from built-in anodes, and inspect the anodes for damage.

☞ Electrical wires should never be used as guidelines or lifting devices.

4. Where anodes are not built-in, powder weld connections, such as Cadweld or Thermite, should be used when safety factors permit.

5. If mechanical connections are necessary, care must be exercised to insure tight, waterproof connections.

6. Carefully route built-in test wires out of the excavation to a convenient, permanent test station.

7. Test for continuity between anode and test wire.

☞ Never use a mechanical compactor on or around built-in anodes.

8. When the backfill reaches the level of the anodes, activate the anodes by pouring 5–10 gallons of water over the anodes.

   Note: It is important to provide a low resistance path for current to leave the anode. If the soil resistance is too high the effectiveness of the cathodic protection is diminished.

☞ All metallic components must be electrically isolated from each other.

9. All cathodically protected components must be isolated from each other.

10. There should be no electrical continuity between tank and piping, tank and fillpipe, or tank and vent line.

11. Steel piping must be protected separately from tanks. Anode design for a piping network must be designed by a certified corrosion engineer.

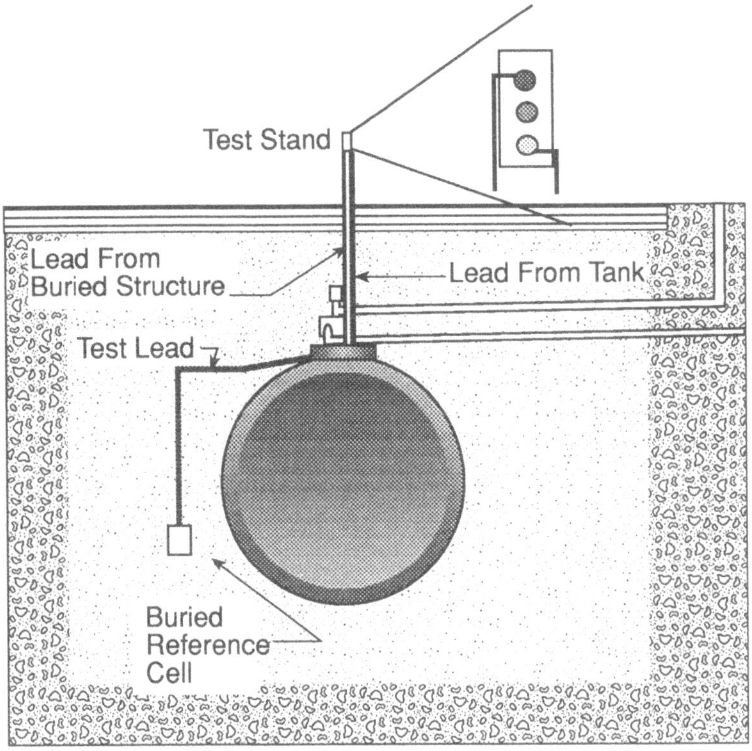

Figure IX–4. Evaluating Cathodic Protection. *Cathodic protection for buried structures is adequate (the anodes are working) if potential between the buried structure and the soil is at least -0.85v relative to a copper–copper sulfate reference cell. Permanently buried reference cells make periodic testing much easier.*

12. Use isolation bushings in all metallic unions.

13. Use a suitable dielectric material as specified by the manufacturer to insulate the tank from hold-down straps.

☞ The structure-to-soil potential is – 0.85 volts relative to a copper–copper sulfate reference cell.

14. If the tank-to-soil potential is – 0.85 volts relative to a copper/copper sulfate reference cell, the tank is protected. If the buried steel structure is *negative* with respect to its environment, its protection

is working regardless of means of protection *(Figure IX–4)*.

15. Permanent copper sulfate or zinc reference cells may be installed within a few inches of the structure, and at a point distant from attached anodes. Test leads from the reference cells are routed out to a test stand.

16. Where sacrificial anodes have been installed, proper operation should be confirmed within 6 months after installation, and every three years thereafter.

Impressed Current/Cathodic Protection

Impressed current systems act to reverse the flow of current in a naturally-occurring corrosion cell by imposing a direct current from an external power source on the components to be protected. Impressed current systems are analogous to a battery that is being recharged. (In the context of battery operation this is called an *electrolysis cell;* see Appendix A).

An external AC-current source is attached to the tank through a rectifier.

☞ If the positive lead is mistakenly attached to the tank, the tank is made the anode and rapid corrosion will result.

- The *negative* terminal of the rectifier is attached to the tank, which becomes the cathode.
- The positive terminal is connected to the non–sacrificial anodes, which are buried metallic electrodes spaced at intervals around the tank.
- Current flows from the tank through the rectifier to the anodes.
- Return current flows through the soil electrolyte from the anodes to the tank (thc cathode).

The effect of the current flow is to maintain the tank at a negative potential with respect to its environment; i.e., the surrounding soil or backfill.

☞ Impressed current systems are widely used to protect long-distance natural gas and crude petroleum transportation pipelines.

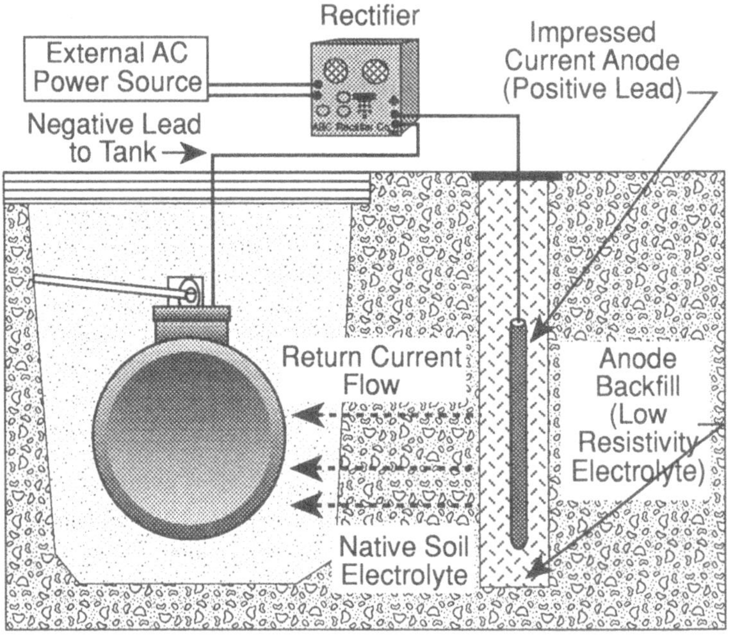

Figure IX–5. Impressed Current Cathodic Protection. *Impressed current/ cathodic protection is an alternative means of corrosion protection, either as original equipment or as a retrofit after anodes have become exhausted. The impressed current configuration has the same component requirements as a galvanic cell except that an external current is imposed on the tank.*

Impressed current/cathodic protection is often the most economical means of controlling corrosion in existing, unprotected, bare steel installations. Other conditions where impressed current/cathodic protection may be required include the following:

- Installations where built-in anodes are used up and cathodic protection must be retrofitted.
- Installations where adjacent steel structures, such as waste oil tanks, hydraulic lifts, or other buried steel tanks induce stray electrical currents.

- Areas near grounded DC electric power operations, such as subways and electric railways. Stray currents enter low-resistance path of the metal; corrosion occurs at points where the current exits the structure.
- Extensive UST and piping systems. Impressed current systems are able to cover wide areas. Large diameter pipelines for natural gas and petroleum products are typically protected by impressed current systems. It is not unusual to protect miles of piping with a single impressed current installation.

A typical impressed current system has two principal components *(Figure IX–5)*:

- A step–down transformer to reduce the AC supply voltage; and
- A rectifier to change the AC current over to a DC output. Rectifier selection depends on the specific conditions, but the rectifier chosen should have excess capacity to allow for adjustments during the life of the cathodic protection system.

Advantages and Disadvantages

Advantages:

- Large driving potential and protection range; impressed current systems can protect an extensive installation and adjacent structures. For large scale installations, particularly installations with extensive steel piping networks, this type of protection system is an appropriate choice.
- Can be retrofitted to protect otherwise sound bare steel tanks. Retrofit is normally combined with relining.

- Flexible current output. Impressed current systems can handle a variety of situations.
- Relatively inexpensive compared to tank replacement costs.

Disadvantages:

- May cause interference problems. This can be a problem in crowded urban sites.
- May be switched off, either manually or from power outage.
- Requires monitoring and routine maintenance.
- Monthly electricity costs may not be trivial.

Installation of Impressed Current Systems

☞ Anode layout is important to proper operation of impressed current systems.

1. Anodes, usually graphite or high silicon cast iron, must be set in a suitable array, as determined by a corrosion expert, about the tank system determined by the current requirements of the system.
2. Anodes are usually backfilled with a carbonaceous material, such as Coke-Breeze™ which lowers anode-to-soil resistance.

☞ As with built–in anodes, it is important to provide a low–resistance path between the anode and cathode for the return current.

3. The *negative* lead of the rectifier must be attached to the tank.

   WARNING: Accidental attachment of the positive lead to the tank will cause rapid corrosion.

4. Powder welds are the preferred means for connecting the negative lead to the buried structure. However, good mechanical connections may be substituted if safety considerations dictate. It is very important to use all appropriate precautions when attaching leads to tanks containing product.

   Warning: Attachment of anodes by thermite welding to tanks or lines which have held flammable or combustible liquids is prohibited unless the tank and surrounding area are made safe beforehand.

5. All connections must be waterproofed with a suitable insulating material.
6. Carefully inspect all wires and connections for damage before backfilling.
7. The AC power source should be on its own separate breaker, in a separate breaker box, equipped with a lock to prevent it from being accidentally turned off.
8. The entire impressed current system must be thoroughly tested at the time of installation, within 6 months of installation or repair to the UST system, and every three years thereafter. A structure–to–soil and structure–to–structure voltage of -0.85v relative to a copper–copper sulfate reference cell indicates the system is operating properly.
9. Rectifiers must be checked every 30 days to be sure the unit is operating properly. In addition the structure–to–soil and structure–to–structure potentials should be checked on an annual basis.

☞ The structure is protected if the test potential is -0.85 volts, or -1.1 volts relative to a buried zinc reference cell.

Chapter X Spill Prevention and Containment

Spills and Overfills

Since the entire purpose of EPA's final regulations is to prevent petroleum products from entering the environment, release containment and detection is a crucial step in any installation. All existing USTs must meet the EPA standards for new tanks by December 22, 1998.

Since spill and overfill releases are generally due to human rather than equipment error, they are usually small in nature, but in time can amount to significant contamination. Consequently, current spill and overfill containment devices on tanks and piping are designed to prevent the following types of releases:

- Spills resulting from the tanker operator disconnecting the fill hose before it has emptied properly.
- Overfills which occur when the tank capacity is exceeded and liquid escapes through vent lines, fill pipes, etc.
- Spills and overfills from fuel dispensing and servicing practices at the dispensing islands.

Spill Prevention

The most common source of soil contamination in existing installations results from overflow or spills at the fill pipe or at the dispenser island. Therefore, the EPA has mandated that new and existing UST systems have devices to prevent spills and overflow.

Spill prevention is accomplished by the use of catchment basins *(Figure X–1)*, also called containment manholes, at the fill pipe and catchment basins at the dispensing islands. At the present time only the former is required. However, installation of catchment basins at the dispensing islands is a relatively inexpensive, prudent form of spill prevention.

☞ For more information on EPA requirements for spill prevention and deadlines see Chapter II, **"EPA Requirements"**, and Appendix B, **"Deadlines."**

Requirements

All USTs must be equipped with one or more of the following:

- A device that will alert the transfer operator when the tank is no more than 90% full by restricting the flow into the tank or triggering a high product–level alarm;
- A device that will automatically shut off flow into the tank when the tank is no more than 95% full;

  or
- An equivalent device approved by the implementing agency.

Catchment basins should be of ample volume, at least 2.5 gallons, and should have a means of draining retained product back into the tank either manually or automatically.

Ball check valves in vent lines and Stage II vapor recovery lines restrict venting when the tank is 90–95% full. The resulting back pressure reduces the gravity flow rate, shutting off flow from the tanker.

High level valves to restrict or stop flow at 90–95% of tank capacity may be located in the fill pipe, inside or outside of the tank. These valves may operate electrically or mechanically and may be combined with an automatic tank gauge.

High level audible and/or visual alarms may be actuated when the tank is 90–95% full. These alarms may operate electrically or mechanically and may be combined with an automatic tank gauge.

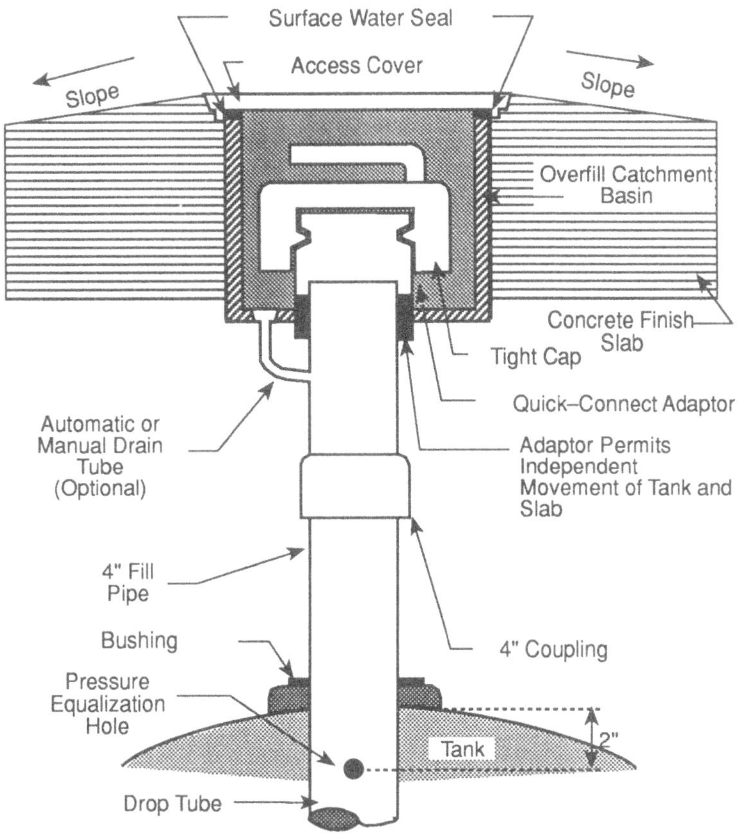

Figure X–1. Fillpipe Spill Containment Basin. *Typical assembly of a small capacity spill containment at the fillpipe. The drain may be manual or automatic. All equipment attached to the tank that goes through the finish slab should be protected from movement by the slab relative to the tank..*

Installation of Catchment Basins

1. Access covers should be installed above grade level to prevent infiltration of surface water.

2. Access covers should be clearly marked with a white triangle and locked to prevent access by the public or inadvertent use as a fillpipe.
3. Since catchment basins may contain flammable vapors, access by the public must be restricted.
4. Follow manufacturer's recommendations for testing, calibration, maintenance, and operation of the particular piece of equipment.
5. Catchment basins at dispenser islands may be installed with their own detection and recovery devices, or be sloped to a piping sump.

Secondary Containment

Secondary containment is defined as complete containment of all possible product release from tanks, piping, dispensers, and other equipment, and is a much more stringent requirement than spill containment. Secondary containment is required on USTs containing hazardous materials or wastes, as defined in CERCLA and RCRA, respectively, but not on petroleum storage tanks. However, many owners and operators are designing secondary containment into new installations, primarily through double-walled equipment, as a prudent business practice, conservative engineering, and good public relations.

Secondary containment devices are designed to meet the following requirements:

- To prevent released product from entering the environment.
- To facilitate monitoring (detection) of product release.
- To make product recovery easier.

There are fives zones in a UST system where secondary containment is practical:

- Tanks and/or the excavation *(Figure X–2).*
  Containment here utilizes double–walled construction or a fully–lined excavation.
- Piping networks.
  The most convenient containment is double–walled piping, although fully–lined trenches are an alternative.
- Dispenser islands.
  Off–the–shelf catchment basins are available. Installation may require modification.

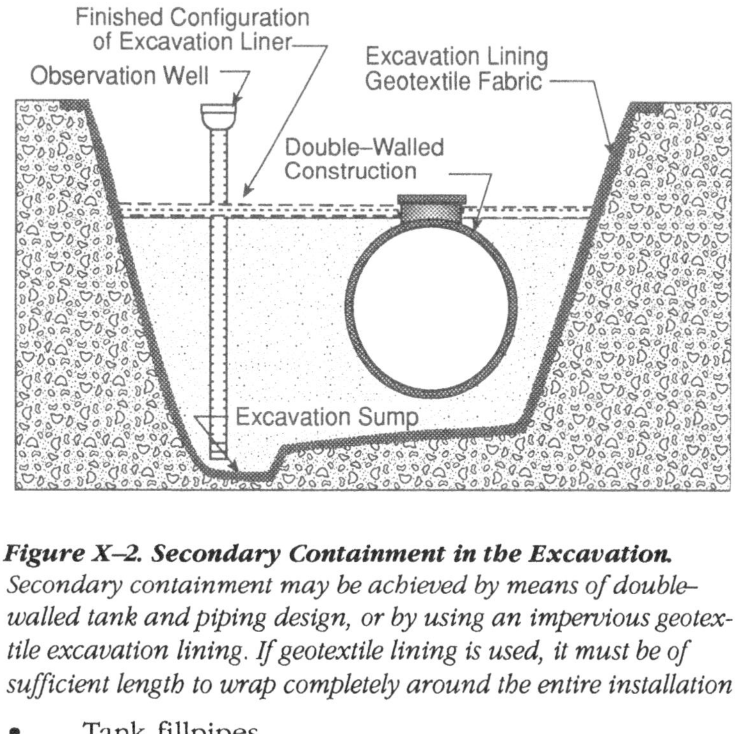

Figure X–2. Secondary Containment in the Excavation.
Secondary containment may be achieved by means of double–walled tank and piping design, or by using an impervious geotextile excavation lining. If geotextile lining is used, it must be of sufficient length to wrap completely around the entire installation.

- Tank fillpipes.
  Catchment basins with manual or automatic drains. These are readily available and required by EPA regulations.
- Submersible pumps.

Catchment basins or partially– or fully–lined trenches are possibilities.

- Vapor recovery at tank and at dispensers. Many states and local authorities require Stage I vapor recovery, and Stage II will be required on higher volume installations by the Clean Air Act of 1990.

Double-Walled Tanks and Piping

- Double–walled equipment is generally the most cost–effective means of containment.
- If double–walled tanks are used, double–walled pipes should also be used.

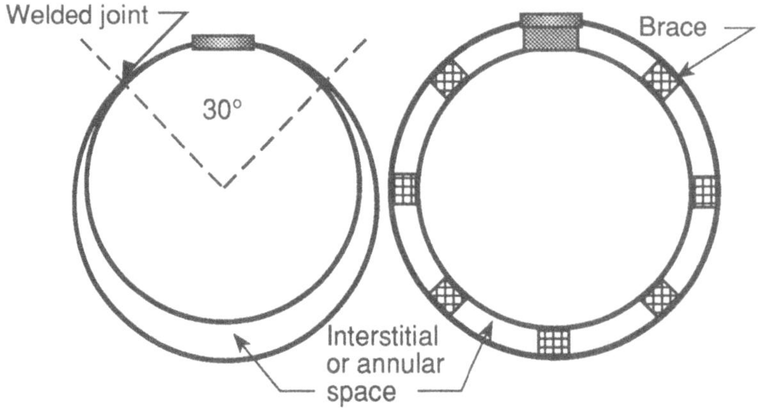

Figure X–3. Double–Walled Secondary Containment. *The most common double–walled design in steel tanks employs an outer jacket that wraps 330° around the inner tank. The design allows for more economic manufacture since the access ports are installed through the single–walled portion. Double–walled FRP tanks typically are a "tank–within–a–tank" design. Since the inner tank is suspended inside the outer shell, structural bracing is required. Either design provides adequate secondary containment and facilitates release detection.*

- Follow manufacturers' recommendations for installation of double–walled equipment.
- As new equipment becomes available, use care in selecting materials compatible with the product to be used, i.e., methanol-containing fuels.

Fully–lined Excavations

Geotextile liners in excavations and trenches are an alternative means of secondary containment. In UST installations where hazardous chemicals are to be stored, geotextile fabric may be required in combination with a clay liner to form an impervious barrier.

In typical UST installations:

- Geotextile liners may be required in installations where methanol will be stored. The installer should consult local regulations.
- Fully lined excavations should be sloped to a sump to facilitate monitoring and recovery.
- Geotextile liners also function as filter fabrics to stabilize excavation walls and prevent backfill migration.

➜ See also Chapter VII, **"Piping and Other Equipment."**

Vent Lines and Vapor Recovery

In addition to preventing spills and overfills occurring during fuel transfer operations, the use of vapor recovery systems is required in many areas in order to reduce hydrocarbon emissions into the atmosphere. Air emissions may not be as readily apparent as soil contamination; nevertheless, in non–attainment urban areas these emissions can further degrade air that is already of poor quality.

Stage I vapor recovery refers to vapor recovery during fuel transfer operations transferring fuel from a tanker to the storage tanks. The most common types of Stage I recovery systems are balance systems, which are collection systems, and vapor processing systems, which is usually some means of disposing of vapors other than collection.

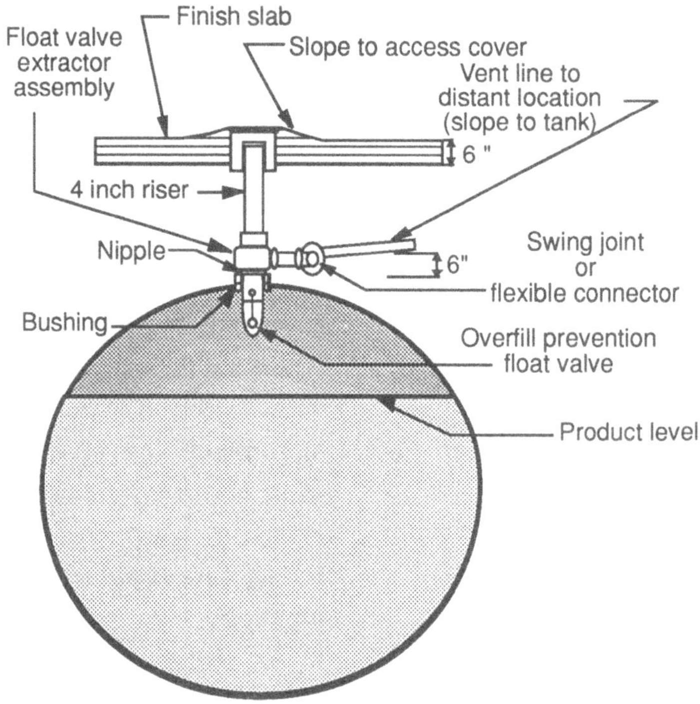

Figure X–4. Vent Line Overfill Protection. *Typical vent line assembly with overfill prevention valve and vapor return riser. The access opening should be located next to the fillpipe opening for convenience. The assembly could be installed inside a small containment basin equipped with an automatic product return for added security.*

The most commonly used vapor recovery method is the balance system, which provides a path for condensed vapors to return to the delivering tank. In the two–point balance system, a vapor return riser is provided for each UST (see *Figure X–4*).

Stage II vapor recovery refers to vapor emission reduction during fuel transfers from dispensers to vehicles (fueling operations at the dispenser islands). Fueling operations are inevitably accompanied by some emission of hydrocarbon vapors , which are oxidized by sunlight, and, thus, increase the burden of carbon monoxide in the atmosphere. Therefore, in urban, non–attainment areas Stage II is required by the Clean Air Act of 1990 at facilities with a throughput of 10,000 gallons per month.

Chapter XI Release Detection Methods

Requirements

Although leak detection methods have been available for many years, EPA requirements have had a profound impact on the industry in the form of new equipment and technologies. Meeting the leak detection requirements is probably the most difficult compliance step for the UST owner/operator.

EPA regulations require that all new and existing UST systems have one or more release detection methods installed that —

- Can detect leaks from any part of the UST system that routinely contains product, including pipes, pumps, or other equipment;
- Is installed, calibrated, maintained, and operated according to the manufacturer's directions;

  and
- Meets the performance standards for new UST systems with any performance claims documented by the manufacturer or the installer.

Any installed release detection method must also monitor the tank at least every thirty days and detect a release with a probability of detection, P_d, of 0.95 and a probability of false alarm, P_{fa}, of 0.5; that is, an alarm or signal indicating a release will have a 95% probability (95 times out of every 100 signals) of being accurate.

By 1998 all existing USTs regardless of age must meet these requirements. An owner/operator of an existing UST system that has spill and overfill containment installed and corrosion protection, if required, may elect to use monthly inventory controls plus tank tightness testing every five years as an alternative release detection method for up to 10 years after installation.

Tank tightness testing must be able to detect a 0.1 gal/hour leak and automatic tank gauges plus inventory control must be able to detect a leak of 0.2 gal/hour. Both new and existing tanks must be monitored at least monthly using one or more of the following methods:

- Automatic tank gauging; *or*
- Vapor monitoring; *or*
- Interstitial monitoring; *or*
- Any other EPA–approved method.

Within these bounds EPA has set no requirements for specific methodology, and new technologies are appearing all the time. Therefore, installers and owners/operators should be aware of the fundamentals of detecting product releases in order to make sound decisions with respect to the most efficient and cost–effective equipment.

Types of Leak Detection

The type of release detection appropriate for a given installation depends on several factors, including:

- Type of tanks: single– or double–walled;
- Soil (backfill) conditions; and
- Groundwater level.

Release detection methods may be conveniently divided into five non–exclusive categories *(Figures X–1 and X–2)*:

- Manual tank gauging, which is limited to smaller–sized tanks, and must be accompanied by tank tightness testing.
- Inventory control, which includes the following:
  - Inventory reconciliation *and*

- — Tank tightness testing.
- Statistical inventory reconciliation.
- External detection, which includes the following:
  - — Soil vapor monitoring;
  - — Groundwater monitoring.
- Internal detection, which includes the following:
  - — Automatic level monitoring;
  - — Interstitial monitoring.
- Pipe release detection, which generally consists of some means of automatic flow restriction device.

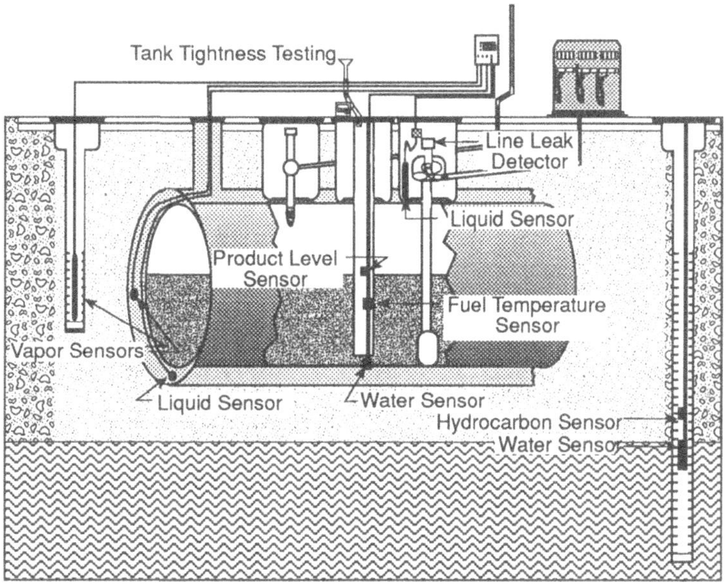

Figure XI–1. Leak Detection in FRP Tanks. *This diagram indicates the various areas of leak detection available for single– and double–walled FRP tanks. From left to right the areas are vapor monitoring, interstitial monitors, in–tank gauges, piping sump and pressurized line detectors, and groundwater monitoring. All of these methods will not be appropriate in any single installation.*

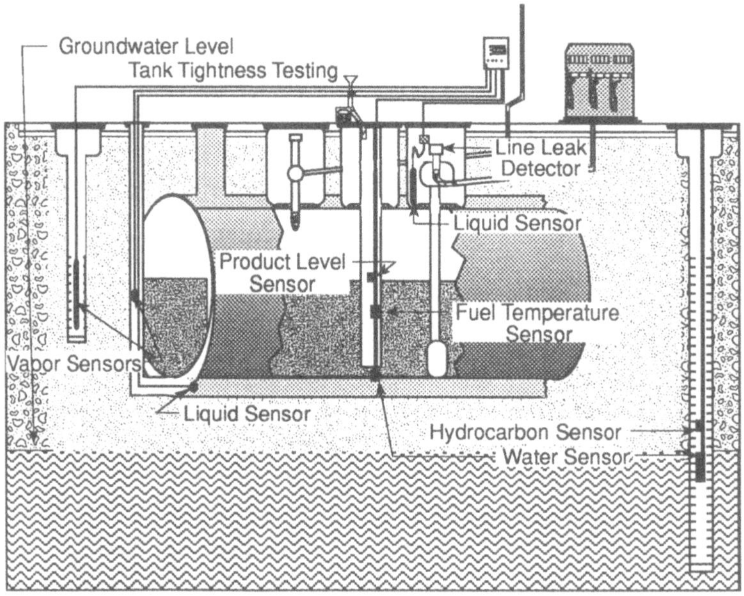

Figure XI–2. Leak Detection in Steel Tanks. *This diagram indicates the various areas of leak detection available for single– and double–walled steel tanks. From left to right the areas are vapor monitoring, interstitial monitors, in–tank gauges, piping sump and pressurized line detectors, and groundwater monitoring. All of these methods will not be appropriate for any single installation.*

☞ Using manual tank gauging requires the UST to be out of service for extended periods of time.

Manual tank gauging is simply measuring the product level in a tank on a weekly basis using a calibrated measuring stick. The linear measurement obtained from the measuring stick is converted to a volume measurement using a calibration chart supplied by the tank manufacturer. This method is one of the least accurate and is limited to tanks of 2,000 gallons or less in two categories.

Operators of tanks of 550 to 1,000 gallon capacity may "stick" a tank on a weekly basis. The tank must not be disturbed for a period of time ranging from 36 to 56 hours before measurement to allow the tank to come to equilibrium. Measuring sticks should be read to the nearest 1/8 inch.

Tanks of 1,000 to 2,000 gallon capacity must combine manual tank gauging with tank tightness testing. Tanks must be measured on a weekly basis.

Inventory control is a basic business practice that, in principle, is simple to implement and cost–effective, and should be an integral part of every UST operation. Inventory monitoring requirements are contained in NFPA 30 and in many state fire codes and UST regulations.

In practice, however, reliance on inventory monitoring alone will not achieve the necessary sensitivity to detect all types of leaks, and must be combined with tank tightness testing. Inventory control should be used in conjunction with more sophisticated technologies, with particular consideration being given to statistical inventory reconciliation.

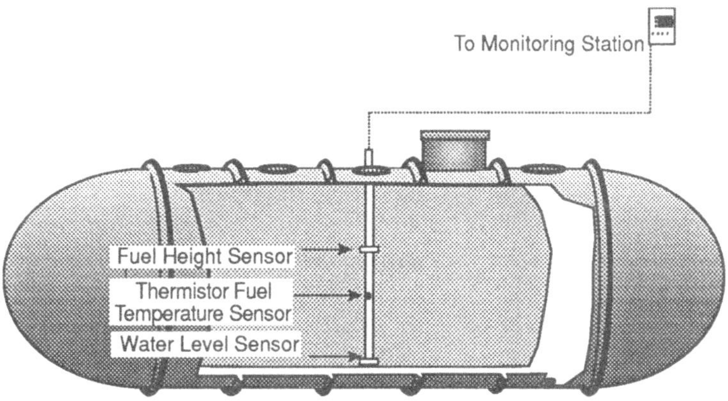

Figure XI–3. Typical Automatic Tank Gauge System. *Automatic tank gauging systems can be very accurate monitors of product volume. An adequate system should have a temperature sensor, product float level, and a water sensor.*

Inventory monitoring in tanks of greater than 2,000 gallon capacity can be done manually by simply measuring the height of product in a UST and converting the measurement to a volume using calibration charts supplied by the UST manufacturer. The measurements must be taken daily, although it is not necessary to take the tank out of service to allow it to come to equilibrium.

☞ Check local regulations for monitoring well specifications.

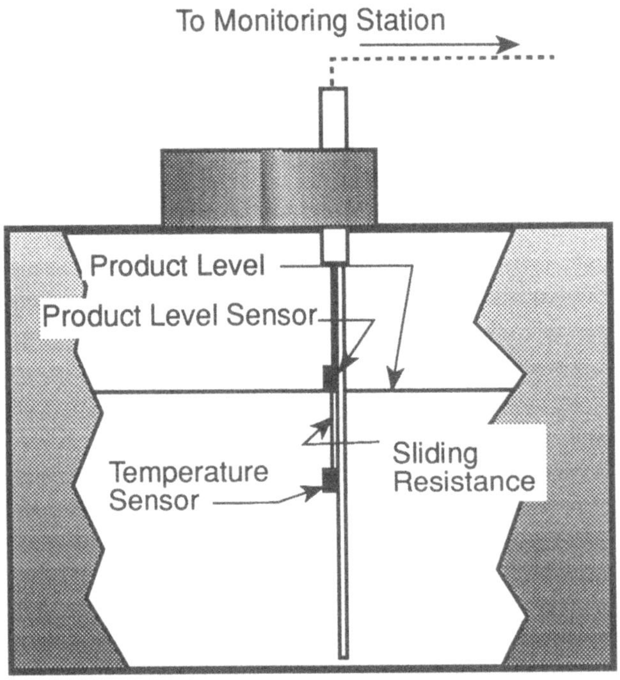

Figure XI–4. Float sensor. *A type of internal tank monitor which can be incorporated into an automatic tank gauge system, in which a float and level sensor (here a simple balance) track changes in liquid level. Signals are converted into electrical signals and sent to a central receiver. The system is not unlike sensors used in gas tanks in automobiles.*

It is a relatively simple matter to send a continuous signal from an automatic tank gauge to a desktop PC; Macs and clones are cheap. If the data were collected in a Lotus or Excel file, then the statistics can be done inside the spreadsheet and graphical display of the inventory is possible. At present no system utilizes this capability, in spite of the fact that people understand and relate to graphical displays far better than numerical displays.

Alternatively, monitoring can be done by any of a number of "off the shelf" automatic tank gauges. An advantage of these systems is that some combine a high level alarm to prevent overfills with volume measurements.

➜ See *"STI Standard for Dual Wall Underground Storage Tanks,"* reference 17.

Internal monitoring systems utilize a variety of devices from simple floats to laser interferometers to measure gain or loss of product or gain or loss of water either in the tank or in the interstitial space. Typical product level sensors are floats that convert resistance to an electrical signal. Product level sensors should take temperature into account since temperature changes can have an appreciable effect on the apparent volume of the tank *(Figures X–3 and X–4)*.

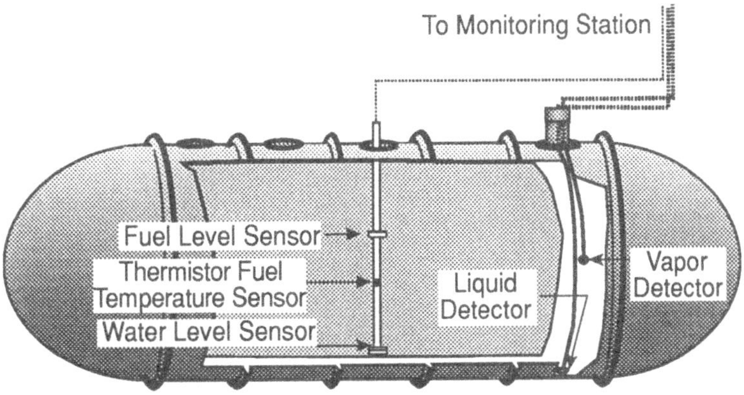

Figure XI–5. Interstitial Monitoring System. *A typical interstitial monitoring system installed in FRP tanks. Adequate monitoring systems should have a temperature sensor, vapor monitor, product sensor and/or water detection. Interstitial monitoring can be very sensitive since the space is quite small and released product can be detected rapidly.*

If the water table approaches the level of a leaking tank, the tank is as likely to admit water as it is to leak product. The reason for this is that the water pressure (pressure head) may exceed the pressure exerted by the product. Therefore, interstice sensors can use product sensitive or water sensitive detectors. Either may be very sensitive to relatively small leaks.

☞ There is no "enter" in "interstice." It is pronounced *in•**ter**•stis.*

Interstice monitoring is a type of internal monitoring available for double-walled steel, FRP, or composite installations. Sensors to detect water and/or hydrocarbons are placed in the interstitial (annular) space between the inner and outer shells. The system may also be equipped with a temperature sensor *(Figure XI–5).*

- Systems are available to continuously monitor either the presence or absence of water or product.
- Detection may be active (designed to signal the presence of specific compounds) or passive (actuate a signal when a current is interrupted).

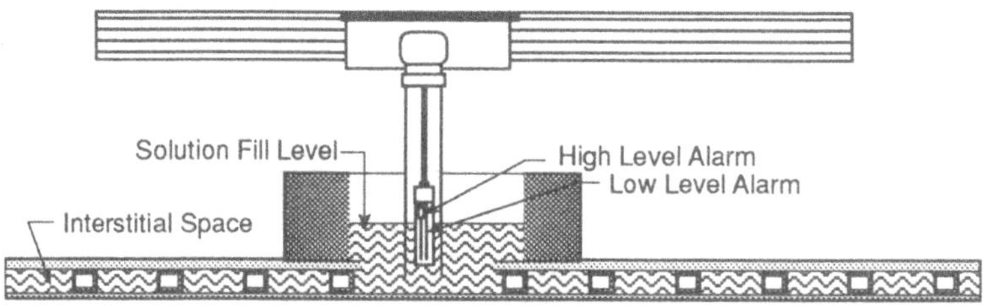

Figure XI–6. Typical Hydrostatic Monitor Assembly. *This type of leak detection was developed by O/C and is a standard practice in European installations. A brine solution is introduced into the interstitial space and drawn up into a narrow neck. A simple float arrangement detects increases or decreases in fluid level, either of which signals a release. The arrangement is elegant, low tech (low maintenance), and capable of considerable accuracy.*

Hydrostatic monitoring *(Figure XI–6)* is a specialized internal monitoring method of leak detection that can be very accurate. The design requires the interstice to be filled with a solution, either a brine solution (a salt water solution containing calcium chloride, $CaCl_2$, in water) or ethylene glycol (antifreeze solution), to prevent freezing. The solution is drawn up into a narrow neck where a fluid level sensor detects changes in volume. The principle is that detecting changes in volume of a liquid in the relatively small volume of the interstitial space is inherently more sensitive than in the much larger volume of the UST itself.

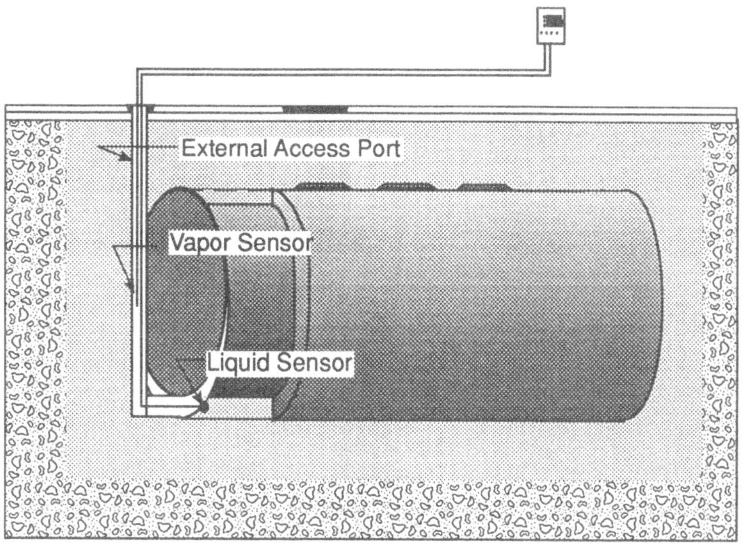

Figure XI–7. Interstitial Monitoring System in Steel Tanks. *Interstitial monitoring systems in steel tanks are similar to those available for FRP tanks. Steel tanks usually have an external monitoring access port built into the shell of the tank. Adequate monitoring systems should have a temperature sensor, vapor monitor, product sensor and/or water detection.*

External monitoring methods utilize observation wells inside the excavation and/or monitoring wells outside the excavation. The wells may contain automatic detection equipment or simply allow access for bailers or other types of sampling.

- **Observation wells** are placed inside the excavation site and typically extend at least 2 feet below the bottom of the tank or into a sump.

- **Monitoring wells** are placed outside the excavation to enable sampling of ground water, and are generally placed downslope from the tank installation.

☞ Be sure to check with local authorities before installing observation wells.

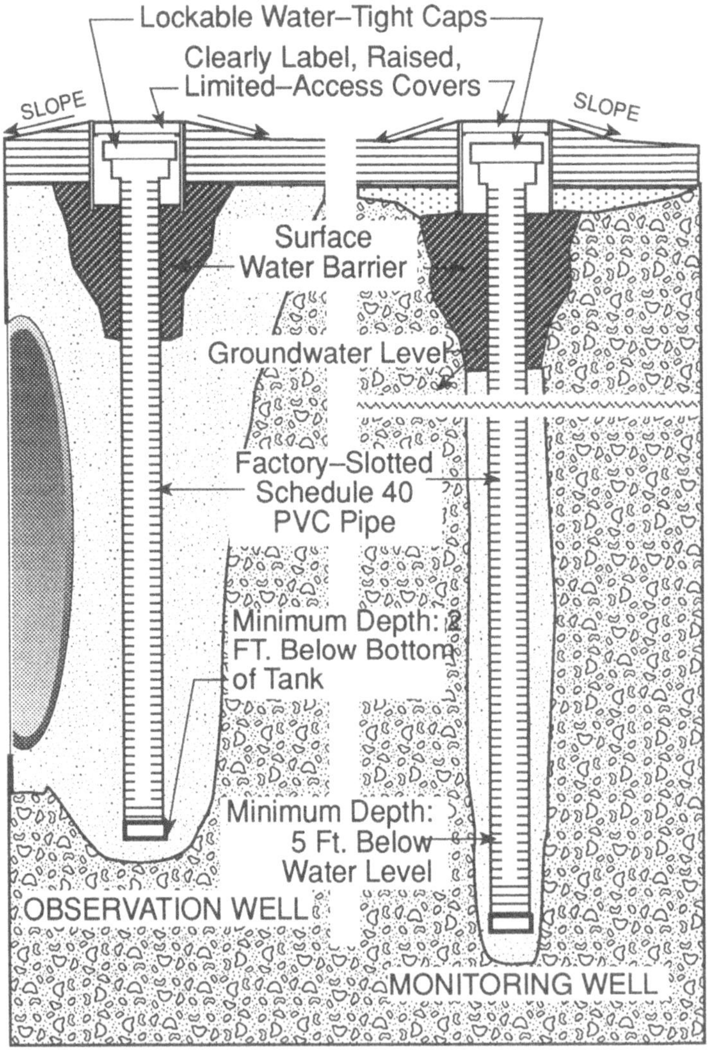

Figure XI–8. Observation and Monitoring Wells. *The construction of the two types of wells is similar; however, monitoring wells are set outside the tank, and generally installed to sample groundwater; observation wells are set inside the excavation to monitor product release.*

Monitoring equipment for external observation wells is similar in operation to that used for internal monitoring. Before using vapor or groundwater monitoring as leak detection, it is necessary to do a site assessment to establish baseline values.

Installation of Monitoring Wells

☞ It is important to restrict entry of surface water. Check local regulations on specific requirements in your area.

1. Placement of wells is strongly dependent on local codes and regulations, and on site-specific conditions. Be sure to check with local authorities for specific regulations.
2. Well bores are typically 10 inches in diameter, and should extend at least 2 feet below the bottom of the tank or hold-down pad. It is usually a requirement that wells extend through the backfill into native soil.
3. If an impervious liner is used, only one observation well is necessary which should extend to within 6 inches of the bottom of the sump.
4. Wells are constructed of perforated PVC, galvanized, or coated metallic pipe with .020–inch openings to permit water and released product to flow into the well.
5. Observation wells must be clearly marked and locked to prevent accidental use.
6. Well bores should be plugged with concrete or bentonite to restrict entry of surface runoff water or product. Monitoring wells should be backfilled with gravel or sand.

Pressurized Line Leak Detectors

All piping that routinely contains product must be monitored for releases. Existing pressurized piping must be equipped with automatic line leak devices that can detect leaks of 3 gallons per hour at 10 pounds per square inch pressure within one hour, and automati-

☞ See Chapter II, "EPA **Regulations**," and Appendix B, **"Deadlines,"** for specifications on suction pumps.

cally shuts off, restricts product flow, and/or triggers an audible alarm if there is a leak.

Existing suction piping must either have a tightness test at least every three years, or be monitored monthly. No release detection is required on certain design requirements.

Line leak detectors are sensitive to pipe pressure loss, and are useful in conjunction in installations with submersible pumps because submersible pumps maintain a constant line pressure. It is possible, and has occurred, that a submersible pump system could continue to pump product out through a leak for an indefinite period of time.

These systems typically may be configured to a variety of functions, depending on installation requirements, including one or more of the following features:

- slow the dispensing rate;
- interrupt electric power to pumps; and/or
- actuate an audible signal.

References

References to the UST literature are surprisingly scattered and diverse. Probably the best single source of information is the Thompson's Underground Storage Tank Guide. The following is a reasonably complete list of pertinent references, listed by organization.

✓ **American Petroleum Institute**
1220 L Street NW, Washington, D.C., 20005. (202) 682-8375.

☞ This is a standard for installation.

1. *Installation of Underground Petroleum Storage Tanks,* API Recommended Practice, **API 1615**, 4th Edition, November, 1987.

☞ This is a standard for removal.

2. *Removal and Disposal of Used Underground Petroleum Storage Tanks,* **API 1604**, Second Edition, 1987.
3. *Recommended Practice for Bulk Liquid Stock Control at Retail Outlets,* **API–1621**, 1989.

☞ This is a standard for assessment and remediation.

4. *A Guide to the Assessment and Remediation of Underground Petroleum Releases,* **API 1628,** Second Edition, August, 1989.
5. *Cathodic Protection of Underground Petroleum Storage Tanks and Piping Systems,* **API 1632**, 2nd Edition, 1988.
6. *Interior Lining of Underground Storage Tanks,* **API 1631**, 1988.

☞ A good collection of six frequently used resources.

7. *Digest of Six Recommended Practices on Underground Petroleum Storage Tank Management,* The preceding six publications are bound together as **API 1650**.

☞ Required reading for installations dispensing methanol blends.

8. *Storage and Handling Of Gasoline–Methanol/ Cosolvent Blends at Distribution Terminals and Service Stations,* **API 1627**, 1986.

9. *Storing and Handling Ethanol and Gasoline–Ethanol Blends at Distribution Terminals and Service Stations,* **API 1626**, 1985.

10. *Waste Water Handling and Treatment Manual for Petroleum Marketing,* **API 1630,** 1984.

11. *Welded Steel Tanks for Oil Storage,* 8th Edition, **API 650**, 1985.

12. *Analysis of Temperature Effects on Gasoline Marketing Operations,* **API 1625**, 1986.

13. *Guidelines for Protection Against Lead Hazard When Dismantling and Disposing of Steel from Tanks That Have Contained Leaded Gasoline,* **API 2202**, 1985.

✓ **Petroleum Equipment Institute**

P.O. Box 2380, Tulsa, Oklahoma, 74101. (918) 494-9696.

☞ This is a standard for installation.

14. *Recommended Practices for Installation of Underground Storage Systems,* **PEI/RP 100-90**, 1990.

✓ **Steel Tank Institute**

728 Anthony Trail, Northbrook, Illinois 60062. (708) 498–1980. Publications available from 570 Oakwood Road, Zurich IL 60047, (708) 438–TANK.

☞ This is the standard for steel tanks by the people who know about steel tanks.

15. *sti–P3 Specification and Manual for External Corrosion Protection of Underground Steel Storage Tanks,* **sti–P3-87**. (Revised, March, 1989).

16. *sti–P3 Installation Instructions & Checklist for Underground Steel Storage Tanks with sti–P3 Corrosion Control System,* **R821–88**, 1987.

17. *Steel Tank Institute Standard for Dual Wall Underground Steel Storage Tanks,* **F841-88**, 1987.

18. *Recommended Practice for Corrosion Protection of Underground Piping Networks Associated With Liquid Storage And Dispensing Systems,* **R892-89**, 1986.

19. *Recommended Practice for Electrical Isolation of Hold Down Straps on Underground Steel Storage Tanks,* **R891–89**, 1984.

20. *Steel Tank Institute Specification for External Corrosion Protection of FRP Composite Steel Underground Storage Tanks,* **F894–89**, 1989.

✓ National Association of Corrosion Engineers

P.O. Box 218340, Houston, TX 77218, (713)492-0535.

☞ Most of what you would need to know about corrosion in buried steel structures.

21. *Control of External Corrosion on Underground or Submerged Metallic Piping Systems,* NACE Recommended Practice **RP–0169**, Item #53002, 1983.

22. *Control of External Corrosion on Metallic Buried, Partially Buried or Submerged Liquid Storage Systems,* NACE Recommended Practice **RP–0285**, Item # 53056, 1985.

23. *Measurement Techniques Related to Criteria for Cathodic Protection of Underground or Submerged Steel Piping Systems as defined in NACE Standard* **RP–0169–83**. Technical Committee Report, T–10A, February, 1990.

24. *Design, Installation, Operation, and Maintenance of Impressed Current Deep Groundbeds,* NACE **RP–0572–85**, 1985.

✓ **National Fire Protection Association**
Batterymarch Park, Quincy, Massachusetts, 02269-9101. (800) 344-3555.

☞ The granddaddy of all standards.

25. *Flammable and Combustible Liquids Code,* **NFPA 30**, 1987.

26. *Automotive and Marine Service Station Code,* **NFPA 30A**, 1987.

27. *Underground Leakage of Flammable and Combustible Liquids,* **NFPA 329**, 1987.

28. *Basic Classification of Flammable and Combustible Liquids,* **NFPA 321**, 1987.

29. Underground Leakage of Flammable and Combustible Liquids, **NFPA 329**, 1988.

✓ **National Leak Prevention Association**
4090 Rosehill Ave. Cincinnati, OH 45229, (800) 543–1838 or (514) 281–7693.

☞ Recommended practices for tank relining, which is not covered in this book.

30. *Spill Prevention, Minimum 10 Year Life Extension of Existing Steel Underground Tanks by Lining Without The Aid of Cathodic Protection,* **NLPA Standard 631,** 1988.

✓ **Western Fire Chiefs Association**
5360 South Workman Mill Road, Whittier, CA 90601. (213) 699–0124.

31. *Uniform Fire Code,* 1988.

✓ **Underwriters' Laboratories, Inc.**
333 Pfingsten Road, Northbrook, Illinois, 60062-2096. (708) 272-8800.

☞ This is another basic standard for steel tanks.

32. *Steel Underground Tanks for Flammable and Combustible Liquids,* **UL–58**, 1986.

33. *Glass–Fiber–Reinforced Plastic Underground Storage Tanks for Petroleum Products,* **UL–842**, 1989.

✓ **Underwriters' Laboratories of Canada**
7 Crouse Road, Scarborough, Ontario, Canada, M1R3A9, (419) 757–3611.

☞ This is a basic standard. It is similar to UL-58.

34. *Standards for Steel Underground Tanks for Flammable and Combustible Liquids,* **CAN/ULC–S603,** 1988.

35. *Standard for Galvanic Corrosion Protection Systems for Steel Underground Tanks for Flammable and Combustible Liquids,* **CAN/ULC–S603.1**, 1985.

✓ **Thompson Publishing Group**
Publication Services Division, 1204 Old Ocean City Road, Salisbury, MD 21801. 1-800-424-2959.

36. *Underground Storage Tank Guide,* Jeffrey L. Leiter, Editor-in-Chief, updated monthly.

☞ These people publish *good* books.

✓ **Lewis Publishers, Inc.**
121 S. Main Street, P.O. Drawer 519, Chelsea, MI 48118. (313) 475–8619.

37. *Underground Storage Systems, Leak detection and monitoring,* Todd G. Schwendeman and H. Kendall Wilcox, 1987.

38. *Groundwater Remediation and Petroleum,* David C. Noonan and James T. Curtis, 1990.

39. *Petroleum Contaminated Soils,* Volume III, Paul T. Kostecki and Edward J. Calabrese, editors, 1990. (See also volumes I, 1988, and II, 1989).

40. *Remedial Technologies for Leaking Underground Storage Tanks,* Lynne M. Preslo, et al., 1988.

Appendix A Batteries and the Corrosion Reactions of Iron

Understanding Corrosion Processes

Buried steel structures rust when the metal is in contact with moist earth because iron releases stored chemical energy in a manner that is exactly analogous to chemical processes that occur in an ordinary dry cell battery. Indeed, a corrosion cell is nothing more than an electrochemical cell set up between the buried structure and a dissimilar reactive material. Therefore, corrosion control is simply a matter of using the principles of battery design to advantage.

➜ See also Chapter IX, **"Corrosion Protection."**

In a battery two or more chemical compounds react spontaneously to produce an electrical current. In buried steel structures iron is spontaneously converted to rust (iron oxide). Corrosion engineers refer to this spontaneous reaction as the tendency of a processed metal to return to its native state.

Batteries

A battery is simply a device to exploit the stored potential chemical energy differences between two different metals, each of which spontaneously tries to return to its natural state. That is, different metals have different reduction potentials. When two different metals, such as iron and copper, are in direct contact, current flows from one to the other because the difference in reduction potentials is expressed as a voltage drop between the two metals.

☞ Flashlight batteries and corrosion cells operate according to the same basic principles.

Galvanic Cells

To construct a usable battery, it is necessary to connect the two metals, and immerse each in an ionic solution. For simplicity suppose the iron electrode is placed in a dilute sulfuric acid/iron sulfate solution. The solution contains hydrogen ions, H^+, iron(II) ions, Fe^{2+}, and sulfate ions, SO_4^{2-}. The copper electrode is placed in a dilute sulfuric acid/copper sulfate solution. This half of the battery solution contains hydrogen ions, H^+, copper(II) ions, Cu^{2+}, and sulfate ions, SO_4^{2-}. The battery will resemble the configuration shown in *Figure A–1.*

☞ **Oxidation** and **corrosion** mean the same thing — and, for iron, both have the same meaning as **"rusting:"**
Metal → Oxide

☞ **Reduction** is the opposite of oxidation. In metallurgy metal ores are **reduced** to produce the pure metal.

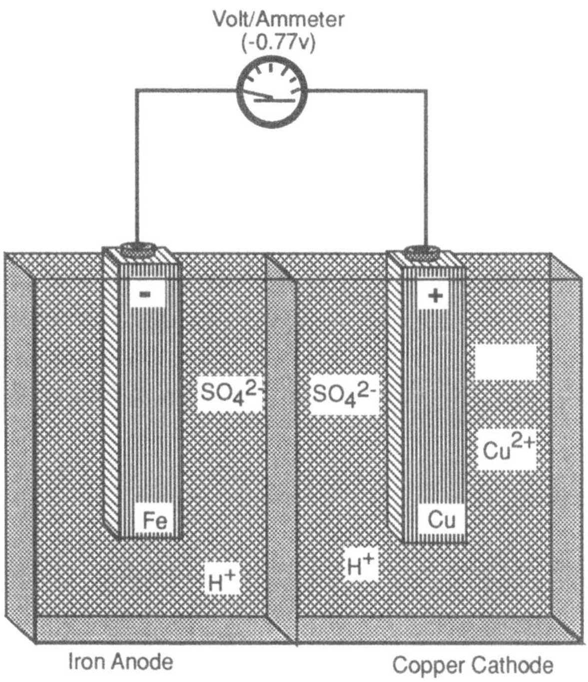

Figure A-1. Components of a Battery. *To utilize the potential difference between two dissimilar metals, it is necessary that they be in direct contact. In the configuration shown in the figure, the two metals are connected through an ammeter; a current will flow briefly then stop since there is no return path to complete the circuit.*

As the battery operates, the iron electrode is consumed since the iron anode is oxidized to iron ions:

$$Fe \rightarrow Fe^{2+} + 2e^- \text{ (oxidation)}$$

The electrons produced from each iron atom force this electrode to be negative and produce the current that flows between electrodes.

☞ Galvanic cells produce energy spontaneously.

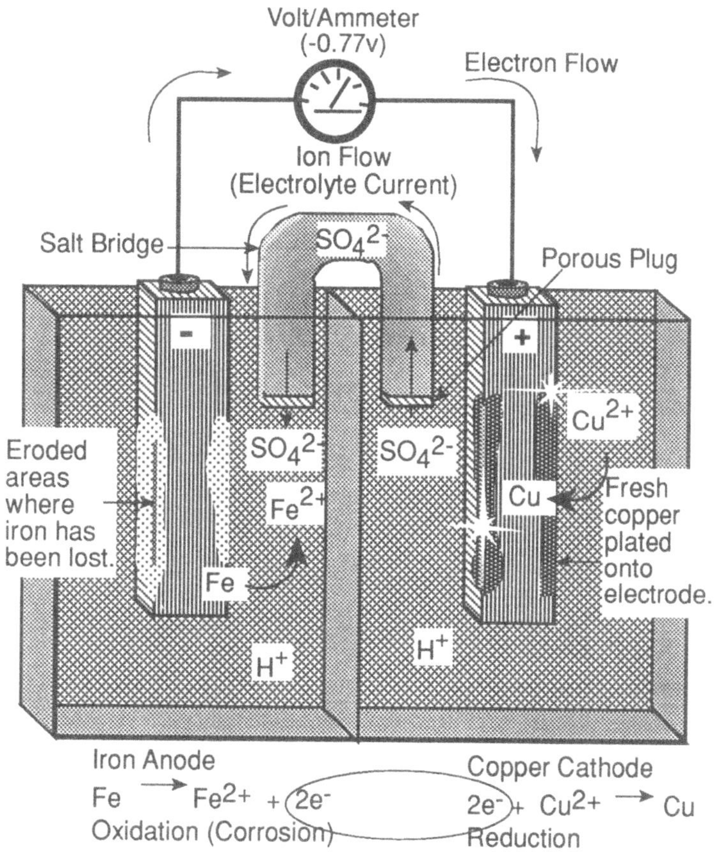

Figure A–2. A Galvanic Cell. *Addition of a salt bridge to the battery in Figure A–1 completes the circuit and allows ions to flow between half cells to equalize the build-up of electrostatic charge. Note that copper from the solution has plated onto the electrode; meanwhile, the anode is losing material through oxidation. This battery will run until the ionic solution is exhausted or the anode is consumed.*

At the other electrode, copper will be deposited:

$$2e^- + Cu^{2+} \rightarrow Cu \text{ (reduction)}$$

☞ In all batteries:
Cathode = Reduction
and
Anode = Corrosion.

Notice that the electrons produced at one electrode are consumed at the other electrode; in the process, current flows through the ammeter and work is done by the battery. This is why the current flow through the metallic connection is important in battery applications; that is where the work is done. In corrosion cells the focus is on the current flow through the soil — the ionic current since that is where corrosion occurs.

☞ The four essential components of any battery are:
- Anode;
- Cathode;
- Metallic connection;
- Electrolyte.

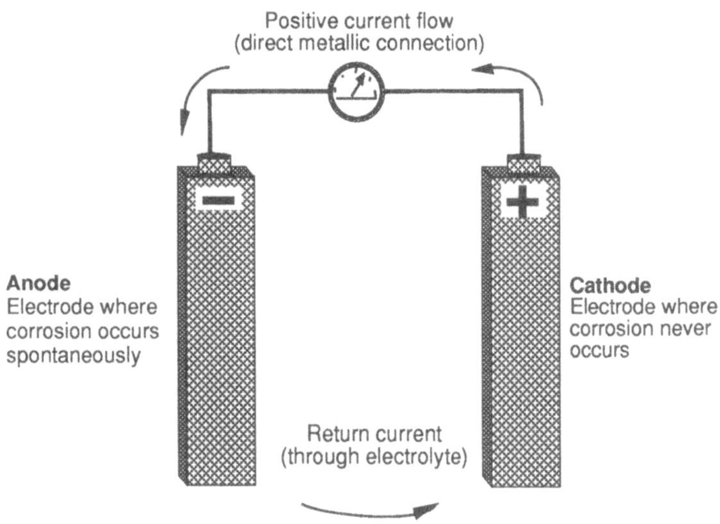

Figure A–3. The Four Essential Components of a Battery. *The four components required to operate a battery are anode, cathode, direct metallic connection, and electrolyte solution to complete the circuit. With these four components, a reversible battery may be operated as either a galvanic (spontaneous) cell or an electrolytic (nonspontaneous) cell. The electrolyte solution completes the circuit by "ionic conductivity" which, in this book, is called the return current.*

In the language of batteries, the electrode where oxidation — or corrosion — occurs is called the *anode;* the electrode where reduction occurs is the *cathode*. Since the object of this discussion is corrosion, henceforth, we will use corrosion instead of oxidation.

As the battery operates the anode solution becomes increasingly positive as Fe^{2+} ions are produced *(Figure A–2)*. Similarly, the cathodic solution becomes increasingly negative as Cu^{2+} ions are used up. The battery becomes polarized, and since the current flow is spontaneous, the battery will operate only briefly until the polarized potential exceeds the chemical potential.

☞ In real batteries, the salt bridge may be nothing more than paper saturated with a salt solution. The purpose is simply to provide a conductive pathway.

A salt bridge is necessary to provide a path for the ionic current — the return current — to complete the circuit. Ions are free to flow between half cells to maintain charge neutrality. In this case negative sulfate ions flow through the salt bridge from cathode solution to anode solution to neutralize the build –up of positive charge (Fe^{2+}) at the anode.

Current flow through the outside circuit must equal the return flow through the electrolyte solution; that is, the electrons (current) produced at the anode must equal the electrons (current) gained at the cathode (circled items in *Figure A–2*).

Thus, there are four essential parts of any battery *(Figure A–3)*:

- An anode where oxidation (corrosion) occurs. This is the negative terminal of a spontaneous battery.
- A cathode where reduction occurs. This is the positive terminal of a spontaneous battery.
- A metallic connection between the anode and cathode. This provides the path for the current to flow between electrodes.
- An electrolyte solution. A conductive solution completes the circuit and provides a path for the

return current. In buried structures the electrolyte solution is moist backfill.

A slightly confusing situation arises with current flow; conventional current flow is defined as a flow of positive charges. In galvanic cells positive current flows through the metallic connection from positive electrode to negative electrode; that is, from cathode to anode. Electron flow is from negative to positive.

The ionic (return) current through the backfill or soil is, by convention, taken to flow from anode to cathode. In buried steel structures corrosion spots are typically observed at anodic areas; that is, points where the positive current flow leaves the surface.

☞ In battery applications the current flow of interest is *cathode to anode* (through the metallic connection) since that is where work is done.

The actual direction of current flow is really immaterial, even DC current. Engineers conventionally speak of positive current flow (flow of positive charges); chemists use negative current flow (flow of negative charges) _ the two are exactly equivalent. It is useful, however, to establish a single convention and stick with it; so current *flows* from anode to cathode.

Electrolysis

A spontaneous battery will produce energy only so long as the electrolyte solution and/or the electrodes last. At some point all batteries will becomes exhausted, unless the battery can be recharged. Recharging means the battery must be run in reverse — in the non–spontaneous direction — until the components of the battery are restored to approximately their original condition. Recharging consists of imposing an external voltage source on the battery, now called an **electrolysis cell** *(Figure A–4).*

☞ An electrolysis cell does not run spontaneously; there must be an external voltage source.

The external voltage must be somewhat greater than the output of the battery in order to cause the chemical reactions to run in reverse. The components of an electrolysis are the same as in a galvanic battery — anode, cathode, metallic connection, and electrolyte, However, the polarity of the electrode is reversed.

☞ Cathode = Reduction and Anode = Corrosion.

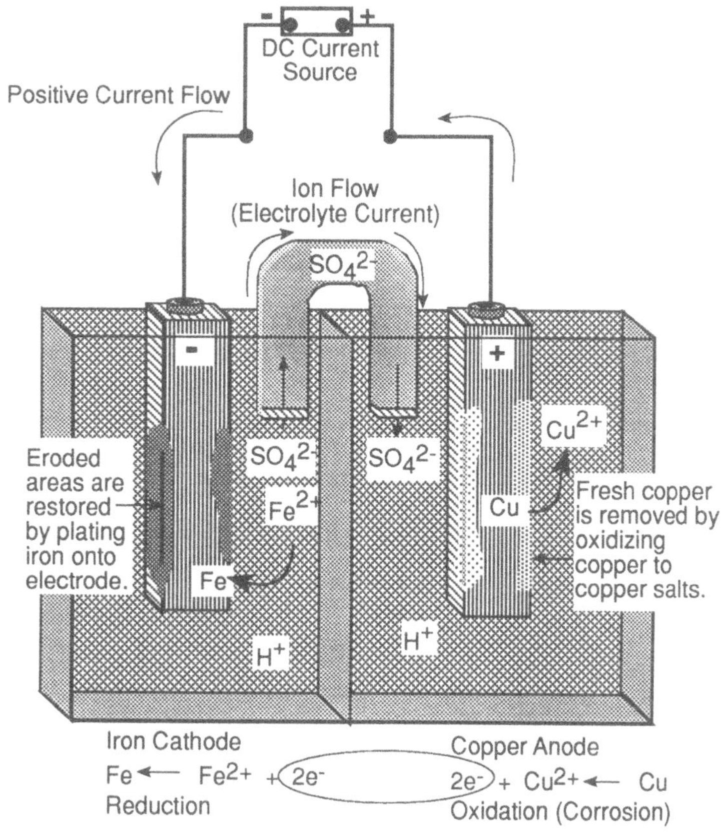

☞ Remember that in all batteries (and buried steel structures) regardless of design:

Cathode Anode
Reduction Corrosion

The only difference is in the polarity.

Galvanic	Electrolysis
Cathode	Anode
Positive	Negative
Negative	Positive

Figure A–4. Electrolysis Cell. *Reversible electrochemical cells can be recharged by imposing an external current source on the cell. Electrons are pumped into the anode to reduce Fe^{2+} ions back to iron, which restores the electrode. Compare this figure with Figure A–2 and note that the iron electrode is still negative, but it is now the cathode. The anode is still the electrode where oxidation occurs, but it is now the positive electrode.*

Sacrificial Anode/Cathodic Protection

Using the copper/copper sulfate half–cell as a reference cell, and measuring the potential generated by various anodic metals, it is possible to construct a table of reduction potentials called a galvanic series. The voltage listed is the potential required to reduce an ion of the metal to the pure metal or alloy, and is proportional to the ease of oxidation of the metal or alloy listed in the table.

☞ The galvanic (or electromotive) series compares the tendency to corrode.

Metal	Voltage
Magnesium (pure)	-1.75
Magnesium alloy (6% Al, 3%Zn, 0.15% Mn)	-1.6
Zinc	-1.1
Aluminum alloy (5% Zn)	-1.0
Mild steel (clean and shiny)	-0.5 to -0.8
Mild steel (rusted)	-0.2 to -0.5
Cast iron	-0.5
Copper/copper sulfate	0.0
Carbon, graphite	+0.3

The utility of this series is that in comparing two metals or alloys, the more negative metal will always be the anode with respect to the less negative, which becomes the cathode. Thus, for example, aluminum alloy (-1.0v) will be anodic with respect to cast iron (-0.5v)

☞ See *Cathodic Protection of Underground Petroleum Storage Tanks and Piping Systems,* API Recommended Practice 1632. Reference 5.

The concept of sacrificial anodes is then straightforward; to protect a buried steel structure simply select a metal more negative than steel and construct a corrosion cell in which steel is the cathode. The anode is allowed to corrode — is sacrificed — to protect the steel structure.

There are many metals suitable for use as anodes, zinc and magnesium are simply the most economical, and most convenient.

☞ This cell has the four basic components: anode, cathode, direct connection, and electrolyte (soil).

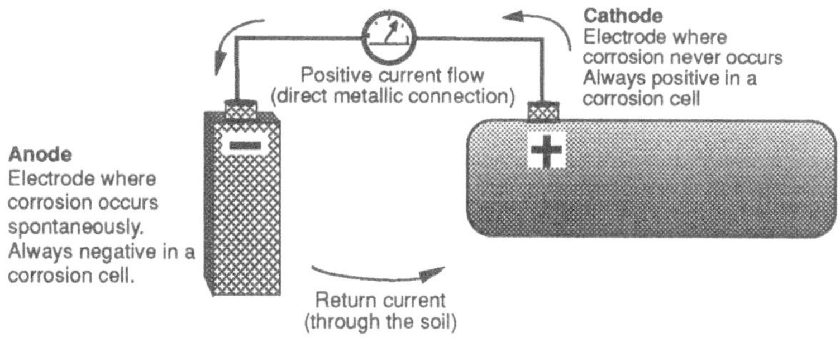

Figure A–5. Sacrificial Anode/Cathodic Protection. *A more negative metal, such as magnesium, is attached to the steel tank. Magnesium becomes anodic with respect to iron and corrodes preferentially, thereby protecting the buried structure. The efficiency depends on the number and placement of anodes and on the resistivity of the backfill. Compare this figure with Figure A–3, p136.*

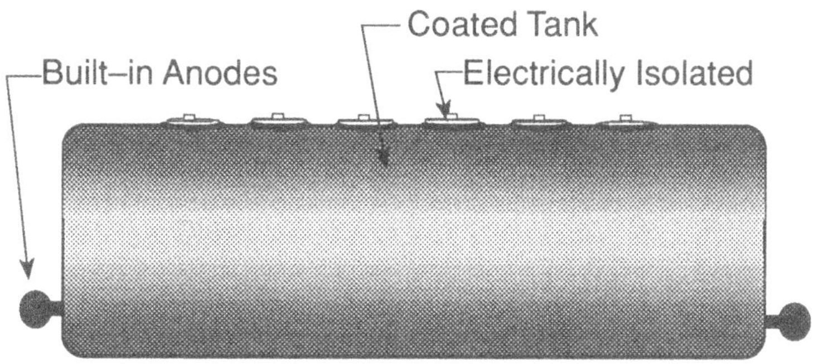

Figure A–6. The STI–P_3 Design. *Steel tanks built according to the Steel Tank Institute, STI, design are shipped with factory-installed anodes. "P3" means three areas of corrosion protection: anodes, coating, and electrical isolation.*

Impressed Current/Cathodic Protection

Impressed current/cathodic protection is similarly straight-forward; it is an adaptation of an electrolysis cell — a corrosion cell is run in reverse. When a buried steel structure is unprotected from corrosion, either due to an original lack of protection or the anodes become exhausted, the spontaneous corrosion cells are unrestricted with the result that the structure begins to rust.

The action of the corrosion cell can be stopped by recharging the battery — that is, by running the cell in reverse. An external voltage source is imposed on the total corrosion cell — the entire UST or other buried steel system. The structure to be protected must be made the cathode and must be negative with respect to the anode just as in the sacrificial anode system.

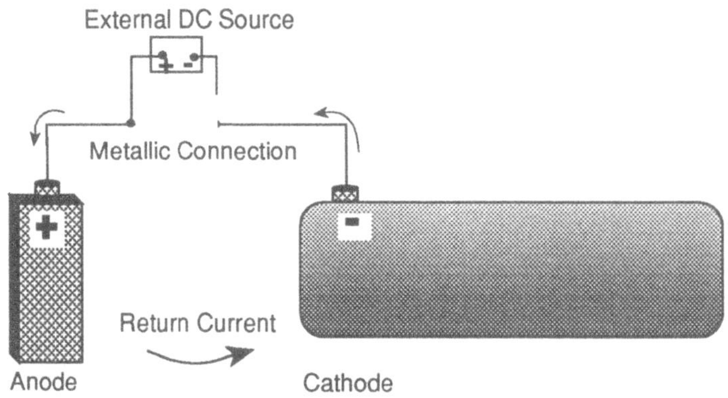

Figure A–7. Essential Parts of an Impressed Current System. *Impressed current systems are identical in principle to sacrificial anode/cathodic protection systems in that the buried steel structure is made the cathode and is still negative with respect to the anode. Note the similarity between this figure and Figure A–5; the required components for an electrochemical cell are all present.*

Rust

The most obvious feature of corrosion in steel structures is the formation of rust, a flaky material not particularly well suited to containing petroleum products. Rust — or, as it is better known, $Fe_2O_3 \bullet nH_2O$ — formation is a complex process in which Fe^{2+} ions undergo further oxidation, precipitation, and polymerization by reaction with environmental factors.

Ferrous iron ions, Fe^{2+}, formed as a result of direct oxidation undergo further reactions with dissolved oxygen to ferric ions, Fe^{3+}:

$$2Fe + O_2 + H_2O \rightarrow 2Fe^{2+} + 4H_2O$$

(Iron) (Ferrous ion)

Ferrous ions are not very stable and react rapidly to the more stable ferric ions:

$$Fe^{2+} + O_2 + H_2O \rightarrow Fe^{3+} + H_2O$$

(Ferrous ion) (Ferric ion)

Ferric ions are insoluble except in very acid solutions and, therefore, form an insoluble, hydrated precipitate:

$$Fe^{3+} + H_2O \rightarrow Fe(OH)_3$$

(Ferric ion) (Ferric hydroxide)

In time the precipitate loses water to form a highly insoluble oxide called "rust":

$$Fe(OH)_3 \rightarrow Fe_2O_3 \bullet nH_2O$$

(Ferric hydroxide) (Rust)

Appendix B Compliance Dates and Deadlines

Compliance Dates for USTs

There are two magic dates for compliance with EPA regulations:

☞ December 22 is simply 90 days after the publication of the EPA Regulations, *Technical Standards.*

December 22, 1988 and **December 22, 1998**

The first determines whether an UST system is new or existing, the second is the final deadline for compliance.

- An existing tank is one that was installed before December 22, 1988;
- A new UST system is one that was installed after December 22, 1988.

For existing UST systems EPA allows owners to phase–in compliance according to a fixed schedule. The rationale is that older systems are more likely to be leaking and thus must be replaced or upgraded first.

This basic schedule is the following ("Installed" refers to Dec. 22 of the year indicated):

Installed	Compliance Date
Before 1965 or Unkown	December 22, 1989
1965 – 1969	December 22, 1990
1970–1974	December 22, 1991
1975–1979	December 22, 1992
1980–1988	December 22, 1993
After 1988	At installation

EPA REQUIREMENTS FOR UST SYSTEMS

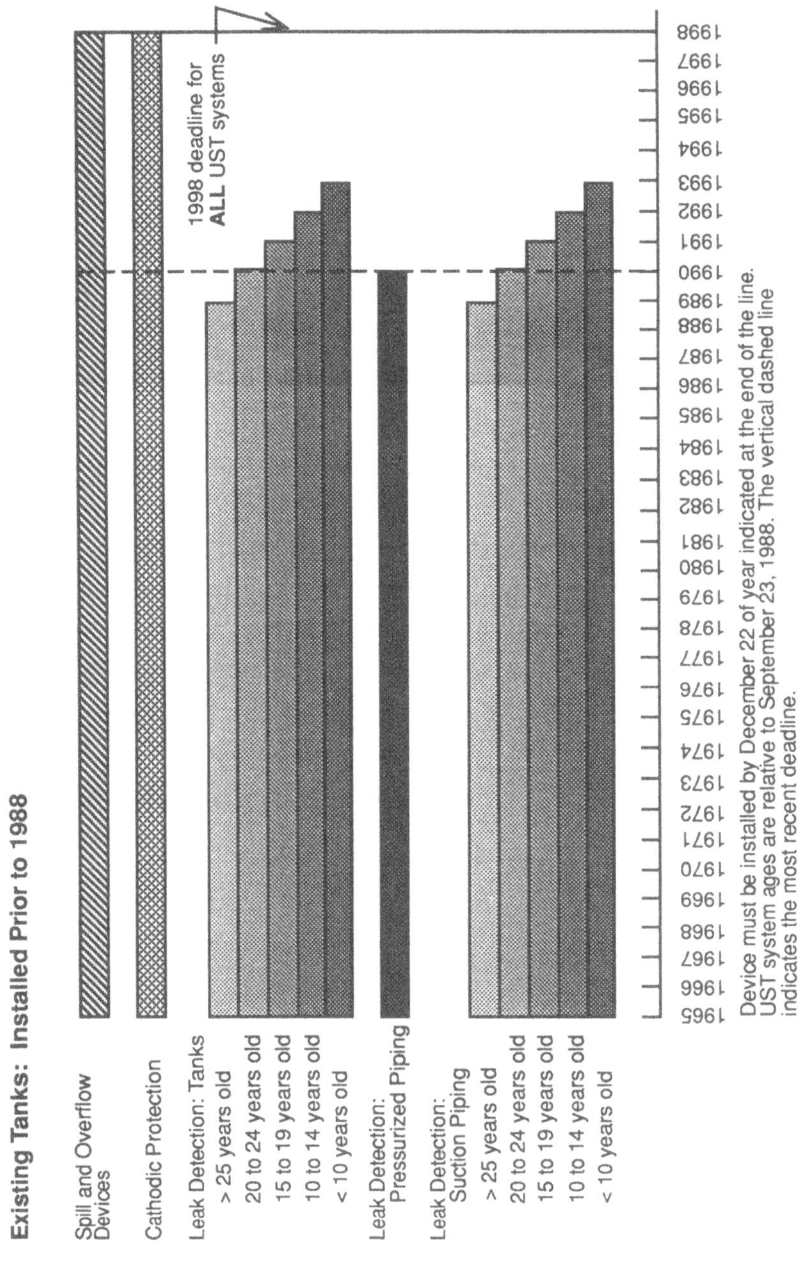

New Tanks: Installed After December 22, 1988

➔ See Chapter IX

Corrosion Protection

➔ All new USTs must be of one of the following designs:
- Coated steel with cathodic protection; or
- Fiberglass; or
- Composite (steel and fiberglass or epoxy).

➔ See Chapter XI

Leak Detection

➔ All new USTs must be equipped for leak detection using one of the following methods:
- Manual tank gauging (550 to 2,000 gallon capacity only), plus tank tightness testing‡; or
- Inventory control, plus tank tightness testing‡; or
- Automatic tank gauging; or
- Vapor monitoring; or
- Interstitial monitoring; or
- Groundwater monitoring; or
- Statistical inventory reconciliation.

‡ New UST systems, assuming they are properly installed with the required spill prevention and corrosion protection, must be tested every five years for 10 years following installation (in principle this can continue until 2008), then use one of the monthly monitoring methods.

➔ See Chapter X

Spill and Overfill Protection

➔ All UST systems installed after Dec. 22, 1988 must be equipped with devices to prevent accidental releases of small amounts of product. Devices include some combination of the following:

☞ Note the "and" in this list of requirements. ☞

- Catchment basins;
  and
- Automatic shutoff devices; or
- Overfill (high level) alarms; or
- Ball float valves.

New Piping: Installed After December 22, 1988

→ See Ch IX

Corrosion Protection

→ All new piping networks must be of one of the following designs:

- Coated steel with cathodic protection; or
- Fiberglass; or
- Other EPA approved design.

Leak Detection

☞ **Also check applicable state regulations.**

→ All pressurized piping must be equipped with some combination of the following devices:

Line leak detection (LLD), which includes —

- Automatic flow restrictor; or
- Automatic shutoff device; or
- Continuous alarm system; or
- Other EPA–approved method;
  and

One of the following —

- Annual line testing; or
- Monthly monitoring (except with automatic tank gauging‡).

→ Suction piping networks require no leak detection devices provided:

- The piping slopes so product can drain back into the tank when suction is released; and

- There is only one check valve which is located as close to the pump in the dispensing unit as possible.

➔ Suction piping networks that do not meet the above design criteria meet be provided with one or more of the following devices:

- Monthly monitoring (not with automatic tank gauging‡), which may include vapor, groundwater, or interstitial monitoring; or
- Line tightness testing every three years.

Existing USTs: Installed Before December 22, 1988

Corrosion Protection

➔ All existing USTs must be of one of the following designs:

- Coated steel with cathodic protection; or
- Fiberglass; or
- Composite (steel and fiberglass or epoxy): or
- Steel with added cathodic protection; or
- Steel with interior relining, or
- Steel with interior relining and cathodic protection.

Leak Detection

➔ All existing USTs must be equipped or retrofitted with leak detection according to the basic phase–in schedule. Leak detection may consist of some combination of the following:

Monthly monitoring, which includes —

- Automatic tank gauging; or

☞ Groundwater or vapor monitoring require a site assessment to establish baselines.

- Vapor monitoring; or
- Interstitial monitoring; or
- Groundwater monitoring; or
- Statistical inventory reconciliation; or
- Other EPA–approved method.

or

- Inventory control, plus tank tightness testing‡; or
- Manual tank gauging (550 to 2,000 gallon capacity only), plus tank tightness testing‡.

‡ Inventory control plus tank tightness testing is an available option only for a 10–year period. Existing UST systems must have a tank tightness test at intervals that depend on the upgrade condition of the tank. There are three possible cases:

- If corrosion protection and spill/overfill prevention are installed on the existing UST **before** December 22, 1988—

  Every five years until Dec. 22, 1998, then go to monthly monitoring.
- If corrosion protection and spill/overfill prevention are installed on the existing UST **after** December 22, 1988—

  Every five years for ten years after upgrade, then go to monthly monitoring.
- If corrosion protection and spill/overfill prevention is **not** installed —

  Every year, then replace or upgrade the UST system by December 22, 1998.

Appendix C Financial Requirements for Owners of UST Systems

Requirements of Subtitle I

The following is typical scenario that, unfortunately, has been enacted repeatedly all across the country. The owner of an UST system decides to install monitoring wells, or he elects to remove the existing tanks and piping and upgrade his system. Whatever the reason, the owner discovers that petroleum contamination exists on his site. It may have been a slow release over a period of many years, accumulated spills due to overfilling the tank, or even a plume from offsite that has migrated onto the owner's property. The origin really doesn't matter, the site must be cleaned up — at an estimated cost of $120,000, including a new UST system.

The owner goes to the bank where he has done business for the past 20 years to borrow a portion of the money to finance cleanup and replacement using the property and facilities to secure the note. The bank asks for an environmental audit of the property. Since the property is contaminated, the bank declines the loan. Now what?

In far too many instances, the owner simply declares bankruptcy and walks away from the property, leaving contamination, cleanup responsibility, liability, tanks, everything. Eventually, the property passes to the state for non–payment of taxes.

In this scenario everyone has lost; the owner has lost his livelihood, the bank lost his business, the state lost revenue and may have to clean up the site. This is not an imaginary scenario. Even major oil companies are being forced to let marginal operations go because the cost of UST system replacement is too high.

As a result, under Subtitle I of SARA, the EPA has established regulations requiring owners of UST systems to demonstrate evidence of financial responsibility for the costs of corrective action and the compensation of third parties for bodily injury and property damage caused by sudden and nonsudden accidental releases arising from operating petroleum USTs. The message is clear — owners with inadequate financial resources should not operate a business that might result in environmental contamination.

➔ 40 CFR §280.90 – 280.109.

For petroleum USTs, EPA requires demonstration of financial responsibility by means of any one, or any combination, of the following mechanisms:

- Financial test of self–insurance;
- Guarantee;
- Insurance or risk–retention group;
- Surety bond;
- Letter of credit;
- State–required mechanism;
- State fund or other state mechanism;
- Fully funded trust fund.

At the time of writing, 43 states have passed legislation to establish assurance funds to assist owners and operators to demonstrate financial responsibility, 21 state assurance funds have been approved, and nine state assurance funds have been proposed.

Financial Requirements

SARA establishes a minimum amount of financial responsibility at $1 million per occurrence for USTs at facilities that are engaged in petroleum production, refining, or marketing. The final regulations require annual aggregate amounts to cover all releases from all petroleum USTs owned or operated. The required aggregate levels are set according to the number of tanks owned — $1 million for up to 100 tanks and $2 million for 101 or more. The final compliance date is October 26, 1991.

Insurance Compliance

At the time of writing, EPA data indicate that about 80% of the owners in the 13–99 tank category have found some form of insurance. Many tank owners have assumed coverage under their comprehensive general liability, CGL, insurance. Indeed, there have been a number of claims filed to clean up contaminated sites under the CGL umbrella. So far the courts have generally taken a "pro–insured" attitude.

However, current policies have increasingly stringent restrictive clauses that limit coverage to "sudden and accidental" releases. Ongoing releases, "leaks" are specifically not covered through conventional CGL policies. Insurance companies writing policies to cover potential environmental damage are listed in Appendix V of Thompson's *Underground Storage Tank Guide*.

➔ See Appendix V, Thompson's *Underground Storage Tank Guide*, reference 36.

Appendix D Summary of Federal and State Programs

EPA Regional UST Offices

Region 1
Maine, Vermont, New Hampshire, Rhode Island, Connecticut, Massachusetts

Region 1
JFK Federal Building
Mailcode: HPU–1
Boston, MA 02203–2211
617- 573–9604
FTS 833–1604

Region 2
New York, New Jersey, Virgin Islands

Region 2
Hazardous Waste Programs Branch
26 Federal Plaza
Mailcode: 2AWM–HWPB
New York, NY 10278
212- 264–3384
FTS 264–3384

Region 3
Pennsylvania, Delaware, Maryland, West Virginia, District of Columbia

Region 3
841 Chestnut St. Bldg.
Mailcode: 3HW63
Philadelphia, PA 19107
215-597-7354
FTS 597-7354

Region 4
Kentucky, Tennessee, North Carolina, South Carolina, Georiga, Florida, Alabama, Mississippi

Region 4
345 Courtland St. N.E.
Mailcode: 4WM-GP
Atlanta, GA 30365
404-347-3866
FTS 257-3866

Region 5
Minnesota, Wisconsin, Illinois, Indiana, Ohio

Region 5
230 S. Dearborn St.
Mailcode: 5HR-JCK-13
Chicago, IL 60604
312-886-6159
FTS 886-6159

Region 6
New Mexico, Texas, Oklahoma, Arkansas, Louisiana

Region 6
1445 Ross Ave.
Mailcode: 6H-A
Dallas, TX 75202-2733
214-655-6755
FTS 255-6755

Region 7
Nebraska, Kansas, Iowa, Missouri

Region 7
RCRA Branch
726 Minnesota Ave.
Kansas City, KS 66101
913-551-7651
FTS 276-7651

Region 8
Colorado, Wyoming, Utah, Montana

Region 8
999 18th St.
Mailcode: 8-HWM-RI
Denver, CO 80202-2405
303-293-1514
FTS 330-1488

Region 9
California, Nevada, Arizona, Hawaii, Guam, American Somoa, Commonwealth of Northern Mariana Islands

Region 9
75 Hawthorne St.
10th Floor, H-2-1
San Francisco, CA 94105
415-744-2079
FTS 484-2079

Region 10
Washington, Oregon, Idaho, Alaska

Region 10
1200 6th Ave.
Mailcode: WD-139
Seattle, WA 98101
206-553-0344
FTS 399-0344

State UST Program Offices

This information is current through September, 1991. For state trust fund information contact the LUST address.

UST/LUST Contact
Dept. of Environmental Conservation
P.O. Box 0, 99811-1800
3030 Vintage Blvd.
Juneau, AK 99801
907-465-2630

UST Contact
Dept. of Environmental Management
Ground-Water Branch/Water Division
1751 Congressman W.L. Dickerson Drive
Montgomery, AL 36120
205-271-7986

UST/LUST Contact
Dept. of Pollution Control & Ecology
P. O. Box 8913, 72219-8913
8001 National Drive
Little Rock, AR 72219
501-562-7444

UST/LUST Contact
Department of Environmental Quality
2005 N. Central Ave., Room 300
Phoenix, AZ 85004
602-257-6984

UST Contact
State Water Resources Control Board
Division of Clean Water Program
P.O. Box 944212, 94244-2120
2014 T St.
Sacramento, CA 95814
916-739-4324

LUST Contact
Department of Health
Underground Storage Tank Program
4210 E. 11th Ave.
Denver, CO 80220
303-331-4830

UST Contact
State Oil Inspection Office
1001 E. 62nd Ave., Room A1
Denver, CO 80216
303-289-5643

UST/LUST Contact
Dept. of Environmental Protection
Underground Storage Tank Program
State Office Building
165 Capitol Ave.
Hartford, CT 06106
203-566-4630

UST/LUST Contact
Environmental Control Division
2100 Martin Luther King Ave., S.E.
Suite 203
Washington, D.C. 20020
202-404-1167

UST/LUST Contact
Dept. of Natural Resources & Environmental Control
Underground Storage Tank Branch
715 Grantham Lane
New Castle, DE 19720
302-323-4588

UST Contact
Dept. of Environmental Regulation
Tank Section
Twin Towers Office Building, Room 403
2600 Blair Stone Road
Tallahassee, FL 32399-2400
904-488-3936

LUST Contact
Dept. of Environmental Regulation
Tank Section
Twin Towers Office Building, Room 409C
2600 Blair Stone Road
Tallahassee, FL 32399-2400
904-488-3936

UST Contact
Environmental Protection Division
Underground Storage Tank Unit
3420 Norman Berry Drive, 7th Floor
Hapeville, GA 30354
404-669-3927

UST/LUST Contact
Department of Health
500 Ala Moana Blvd.
Honolulu, HI 96813
808-543-8226

UST/LUST Contact
Department of Natural Resources
Henry A. Wallace Building
900 E. Grand Ave.
Des Moines, IA 50319
515-281-8135

UST Contact
Department of Health & Welfare
450 W. State St.
Boise, ID 83710
208-334-5860

LUST Contact
Division of Environmental Quality
450 W. State St.
Boise, ID 83710

208-334-5860

UST Contact
Office of State Fire Marshal
3150 Executive Park Drive
Springfield, IL 62794-9276
217-782-6761

UST/LUST Contact
5500 W. Bradbury Ave.
Indianapolis, IN 46241
317-243-5022

UST/LUST Contact
Department of Health & Environment
Bureau of Environmental Remediation
Underground Storage Tank Section
Forbes Field, Building 740
Topeka, KS 66620
913-296-1678

UST/LUST Contact
Division of Waste Management
Underground Storage Tank Section
18 Reilly Road
Frankfort, KY 40601
502-564-6716

UST Contact
Dept. of Environmental Quality
Underground Storage Tank Division
P.O. Box 44274, 438 Main St.
Baton Rouge, LA 70804
504-765-0243

UST Contact
Department of Public Safety
Underground Storage Tank Program
P.O. Box 490, East St., Building No. 5
Tewkesbury, MA 01876
508-851-9813

LUST Contact
Dept. of Environmental Protection
1 Winter St.
Boston, MA 02108
617-292-5903

UST/LUST Contact
Department of Environment
Hazardous & Solid Waste Mgmt. Admin.
Underground Storage Tank Section
2500 Broening Highway
Baltimore, MD 21224
301-631-3442

UST/LUST Contact
Dept. of Environmental Protection
State House - Station 17
Hospital Street, Ray Building
Augusta, ME 04333
207-289-2651

UST Contact
Department of State Police
Fire Marshal Division
P.O. Box 30157
Lansing, MI 48913
517-334-7079

LUST Contact
Department of Natural Resources - ERD
P.O. Box 30028
Lansing, MI 48913
517-373-2794

UST/LUST Contact
Pollution Control Agency
Underground Storage Tank Program
520 Lafayette Road N.
St. Paul, MN 55155
612-296-8100

UST Contact
Department of Natural Resources
P.O. Box 176
205 Jefferson St.
Jefferson City, MO 65102
314-751-7428

LUST Contact
Department of Natural Resources
Laboratory Services Program
2010 Missouri Blvd.
Jefferson City, MO 65101
314-751-7929

UST/LUST Contact
Department of Environmental Quality
Bureau of Pollution Control
Underground Storage Tank Section
P.O. Box 10385
2380 Highway 80 W.
Jackson, MS 39289-0385
601-961-5171

UST/LUST Contact
Dept. of Health & Environmental Science
Solid & Hazardous Waste Bureau
836 Front St.
Helena, MT 59701
406-444-5970

UST/LUST Contact
Pollution Control Branch
Division of Environmental Management
Dept. of Env. Health & Natural Resources
441 N. Harrington St.
Raleigh, NC 27603
919-733-8486

UST Contact
Department of Health
Division of Waste Management
Box 5520, 1200 Missouri Ave.
Bismarck, ND 58502-5520
701-221-5166

UST Contact
State Fire Marshal's Office
Underground Storage Tank Division
246 S. 14th St.
Lincoln, NE 68508
402-471-9465

LUST Contact
Dept. of Environmental Control
Groundwater Sect., Water Quality Div.
301 Centennial Mall S., Box 98922
Lincoln, NE 68509-8922
402-471-4230

UST/LUST Contact
Dept. of Environmental Services
Underground Storage Tank Program
6 Hazen Drive, P.O. Box 95
Concord, NH 03301
603-271-3644

UST/LUST Contact
Dept. of Environmental Protection
Div. of Water Resources (CN-029)
401 E. State St.
Trenton, NJ 08625
609-984-3156

UST Contact
Environment Department
Underground Storage Tank Bureau
P.O. Box 26110
1190 St. Francis Drive
Harold Runnels Building, Room N2150
Santa Fe, NM 87502

505-827-0188

UST/LUST Contact
Dept. of Conservation & Natural Res.
Division of Environmental Protection
Capitol Complex
123 W. Nye Lane
Carson City, NV 89710
702-687-5872

UST Contact
Dept. of Environmental Conservation
Bulk Storage Section
50 Wolf Road, Room 326
Albany, NY 12233-3520
518-457-4351

UST Contact
Department of Commerce
7510 E. Main St.
P.O. Box 525
Reynoldsburg, OH 43068
614-752-7938

LUST Contact
Department of Commerce
8895 E. Main St.
Reynoldsburg, OH 43068-3395
614-752-7938

UST/LUST Contact
Corporation Commission
Underground Storage Tank Program
Jim Thorpe Building
2101 North Lincoln Blvd.
Oklahoma City, OK 73105
405-521-3107

UST Contact
Dept. of Environmental Quality
811 SW Sixth Ave., 7th Floor
Portland, OR 97204
503-229-5733

LUST Contact
Dept. of Environmental Quality
UST Cleanup Program
811 SW Sixth Ave., 9th Floor
Portland, OR 97204
503-229-6170

UST/LUST Contact
Dept. of Environmental Resources
Storage Tank Program
3600 Vartan Way, 2nd Floor
P.O. Box 8762
Harrisburg, PA 17105-8762
717-657-4080

UST/LUST Contact
Dept. of Environmental Management
Underground Storage Tank Section
291 Promenade St.
Providence, RI 02908
401-277-2234

UST/LUST Contact
Dept. of Health and Environ. Control
Ground-Water Protection Division
2600 Bull St.
Columbia, SC 29201
803-734-5332

UST Contact
Dept. of Environ. & Natural Resources
Office of Water Quality
523 E. Capitol
Joe Foss Building
Pierre, SD 57501-3181
605-773-3351

UST/LUST Contact
Dept. of Health & Environment
200 Doctors Building
706 Church St.

Nashville, TN 37247-4101
615-741-4081

UST/LUST Contact
Texas Water Commission
Petroleum Storage Tank Division
P.O. Box 13087, 1700 N. Congress
Austin, TX 78711
512-371-6200

UST Contact
Department of Health
Bureau of Solid & Hazardous Waste
Solid and Hazardous Waste Section
P.O. Box 16700
Salt Lake City, UT 84116-0700
801-538-6338

UST/LUST Contact
State Water Control Board
P.O. Box 11143
2111 N. Hamilton St.
Richmond, VA 23230-1143
804-367-6685

UST/LUST Contact
Dept. of Natural Resources
Underground Storage Tank Program
103 S. Main St., West Building
Waterbury, VT 05676
802-244-8702

UST Contact
Department of Ecology
Mail Stop PV-11
Olympia, WA 98504-8711
206-459-6272

LUST Contact
Department of Ecology
Toxic Cleanup Program
Mail Stop PV-11
Olympia, WA 98504-8711
206-438-3000

UST Contact
Dept. of Industry, Labor & Human Relations
P.O. Box 7969
201 E. Washington Ave.
Madison, WI 53707-7969
608-267-9725

LUST Contact
Department of Natural Resources
P.O. Box 7921
Madison, WI 53707
608-267-7560

UST/LUST Contact
Waste Management Section
Division of Natural Resources
UST/LUST Office
1356 Hansford St.
Charleston, WV 25301
304-348-6371

UST Contact
Dept. of Environmental Quality
Water Quality Division
Herschler Building, 4th Floor W.
122 W. 25th St.
Cheyenne, WY 82002
307-777-7781

UST/LUST Contact
Environmental Protection Agency
Office of the Governor
American Samoa Government
ATTN: UST/LUST Program
Pago Pago, American Samoa 96799
684-633-2682

UST/LUST Contact

Division of Env
Environmental Quality
P.O. Box 1304
Commonwealth of Northern Mariana
Islands
Saipan, CM 96950
607-234-6984

Index

Figures are indicated by page numbers in italics.

Symbols

45° Rule 25

A

Abandoned tanks 1
Accessories
  typical *72*
ACT. *See* Association of Composite Tanks
ACT–100 15
Advantages
  impressed current systems 101
  sacrificial anode protection 96
Air pressure gauges 43–44
Alarm 115
American Society for Testing and Materials, ASTM 15
Anchorage
  types 59–61
Anchors 57–59
Annual line testing 20, 148
Anode
  defined 93
  operation 136–137
Anode type 96
Associated equipment 25
Association of Composite Tanks 15
ASTM. *See* American Society for Testing and Materials
Automatic flow restriction device 20, 148
Automatic level monitoring 117
Automatic shutoff device 20, 148
Automatic tank gauges 18–19, *119*
  installation 72
  leak detection 116
  EPA requirements 147, 149

B

Backfill 49–50, 58
  flotation 58
  in corrosion cells 94
  piping trenches 69
  voids 49–50
  with built–in anodes 97
Backfill bedding
  piping 69
  tanks 50–51
Backfill materials
  and corrosion protection 49
  appropriate 49
  equipment 50–52
  FRP tanks 54–55
  pea gravel 54
  steel or composite tanks 54
Backfilling 24
Ball float check valve 20, 148
Ballasting 53–54
Basic compliance schedule 145
Bathtub effect 53–54
Batteries 7
  defined 133
  essential components of a battery *136*
Bedding layer 50
Benzene 11, 83
Bottom hold-down pad 59–61, *61*
Buoyant forces 57–59
Burial depth 27
  flotation 58
  maximum 33
  minimum 32
  piping, minimum 70
Buried steel structures 133
Bushings
  nylon 72

C

Cast iron plugs 42–44
Catchment basins 19, 148
Cathode
  defined 93
  operation 137
Cathodic protection 17
  sacrificial anode 95
CERCLA 9, 10–11
CFR (40 CFR), *See* Code of Federal Regulations
Check valves
  installation 73
Chemical energy 133
Chemical properties
  in steel 93
Chronic effects
  petroleum toxicity 83
Closure 88
Closure in Place
  procedure 89
CO_2 84 *See also* Dry ice
Coated steel 149
Coating 41
Code of Federal Regulations 12
Combustible gas indicator 83
Compaction 24
  inadequate 49
Compliance dates 8, 145
Components of a battery *134*
Composite tanks 41, 149
Comprehensive Environmental Response, Compensation and Liability Act 9
Compressional forces 38–39
Continuous alarm system 20, 148
Copper/copper sulfate reference cell 98, 139
Corrosion 91
  buried steel structures 92
Corrosion cell
  and batteries 133
  planned 95
Corrosion Cell in Dissimilar Soils
  figure 94
Corrosion cells 92–95
Corrosion control 133
Corrosion due to dissimilar metals *92*
Corrosion processes 133
Corrosion protection
  and backfill 49
  requirements 7, 147–149
Corrosion reactions 7
Corrosive soils 96
Critical backfill areas 51–52
  multi tank installations 52
  voids in 49–50
Crushed rock 54
Current flow
  convention 138

D

DC electric power operations 101
De minimus 14
Deadman Anchors 59–61, *60*
  installation 61
Deflection 33
  and tank failure 49–50
Depth of Cover
  dry installations 33
  wet installations 34–35
Design criteria
  buried steel structures 91–92
Dielectric bushings 42, 45
Dielectric material 98
Disadvantages
  impressed current systems 101
  sacrificial anode protection 96
Disposal
  disposal sites 88
  labeling 87

used USTs 86
Dissimilar metals 92
Dissimilar soils 94
Dry cell battery 133
Dry ice 84 *See also* CO_2

E

Electrical continuity 97
Electrode 133
Electrolysis 138
Electrolysis cell *139*
Electrolyte material
built–in anodes 95
Electrolyte solution
defined 93
and soil current 137
Environmental damage 2
Environmental Protection Agency 2, 23, 28, 49, 67, 115
EPA, *See* Environmental Protection Agency
Epoxy resin 93
Equipment
Equipment access opening *79*
associated 57
backfill requirements 50–52
inspection 46
lifting 38–40
Essential parts of an impressed current system *142*
Essential parts of any battery *137*
Ethanol 16
Ethylbenzene 11
Evaluating Cathodic Protection
figure 98
Excavated materials 29
Excavation 6, 23–27
backfill 50
inspection 41
safety 23
tank removal 84
Exclusions
to UST definitions 13
Existing USTs
leak detection 115
requirements 145, 149
Explosive vapors 83
External detection 117
External monitoring 123

F

Farm tanks 13
Fiberglass reinforced plastic 17
Fiberglass resin 93
Fiberglass Tank and Pipe Institute 15
Fiberglass-reinforced plastic 15–16
backfill support 49
hold–down straps 63–64
Fill lines
installation 72
Filter fabric 31, 55–56
Financial requirements 8, 150
Financial responsibility 149
Fittings
soap test 75
Flexible connectors 73, 74
Float sensor
figure 120
Flotation 57–59
burial depth 57–59
Flow–through process tanks 13
Forces Associated with Lifting *39*
FRP. *See* Fiberglass-reinforced plastic
FRP Pipes
installation 74
Fuel additives 16

G

Galvanic anode protection 95
Galvanic Cell *135*
Galvanic cells 133
Galvanic series 140

Gasoline vapors 1
Geotextile liner 31
Graphical display 120
Grease traps 13
Groundwater level 116
Groundwater monitoring 18–19, 117 147, 149
Guidelines 40

H

Handling of Excavated Materials 29
Handling Tanks 37
HASWA 9. *See also* Hazardous and Solid Waste Amendments
Hazardous and Solid Waste Amendments 9. *See also* HASWA
Hazardous materials 9
Hazardous substance 12–14
Heating oil tanks 13
Hold–down Pads
  installation 61
Hold–down straps 60–61
Hydraulic lift oil storage tanks 14
Hydrocarbon toxicity 82
Hydrostatic monitoring 122
Hydrostatic tests
  pipes 75

I

Impact (fire) valves 73
Impressed current/cathodic protection
  requirements 17
  defined 99
  figure *100*
Impressed current system 142
Impressed current/cathodic protection
  operation 141
Inert gas
  in tank removals 84
Inerting 84, 85
Installation
  UST systems 4–5
Installation
  deadman anchors 61–64
  hold–down pads 61–64
  impressed current systems 102
  mid–anchoring 64–65
  metallic piping 73–74
  monitoring wells 125
  nonmetallic piping 74
  observation wells 125
  piping 71
  piping system 67
  sacrificial anode protection 96
  vapor recovery 78
Installation procedures 4
Insurance Compliance 150
Insurance group 149
Internal detection 117
Internal monitoring systems 121
Interstitial monitoring 18–19
  devices 116–117, *121*
  leak detection 122
  in steel tanks 123, *123*
Interstitial space 44–46
Inventory control 116, 119, 147, 150
Inventory monitoring 7, 147, 149
Ionic current 138
Iron
  corrosion reactions 144
Isolation bushings 98

J

Joints
  cutting, reaming 74
Joints and fittings
  soap test 75

L

Laboratory analysis 88

Lead vapors 87
Leaded gasoline 87
Leak detection 7, 18,
  EPA requirements 148
  in FRP tanks *117*
  in steel tanks *118*
Leak detection methods 115–116
Leaks 115
Legislation 6
License (installer) 3
Lifting angle 39–40
Lifting force 40
Lifting lugs 38
Line leak detectors 126, 148
Liquid traps 13
Long term exposure
  petroleum toxicity 83
Lower explosive limit (LEL) 83
LUST trust fund 8
LUSTs 1

M

M85 16
Magnesium anodes 96
Management
  UST systems 4–5, 9
Manufacturer's recommendations 24
Manual tank gauging 116, 118, 147, 150
Manway access ports 65
Manways 73, 79
Material handling 6
Material inspection 46
Material storage 41
Materials 69
Metallic piping
  installation 73–74
Methanol 11, 16
  compatibility 71
Mid–anchoring 60–61
Monitoring equipment 72
Monitoring wells 124
  installation 125
  placement 125
  site assessment 88
Monthly inventory control 18–19
Monthly monitoring 148
Multiple tank installations 62

N

National Fire Protection Association 3
NFPA *See* National Fire Protection Association
New Piping 148
New UST system 145

O

Observation and monitoring wells 124, *124*
Occupational Safety and Health Administration 51–52
Operation
  UST systems 9
Operator 14
OSHA *See* Occupational Safety and Health Administration
Overfill (high level) alarms 148
Overfill containment boxes
  inspection 46–47
Overfill devices 73
Overfill protection 19
Overfills 1, 7
Owner 14
Oxidation 136, 137
Oxygen deficiency 85

P

Pea-gravel 54
PEI *See* Petroleum Equipment Institute

Perforated PVC 125
Performance standards 115
Petroleum Equipment Institute 81
Petroleum fuels
  explosive 81
  flammable 81
  toxicity 82
  symptoms 81
Petroleum toxicity 82
  chronic effects 82–83
  long–term effects 83
Petroleum UST system 14
Pipe
  black steel 73
  FRP 69
Pipe dope 71
Pipe release detection 117
Pipe size 68
Pipe tightness tests 74–75
Pipeline facilities 13
Piping 20, 57
  failures 91
  inspection 41
  installation 71
  tightness tests 74
Piping and fittings
  inspection 46–47
Piping failures 91
Piping Layout 67–68, 68
Piping requirements 25
Piping slope 68
Placement 27
Polarized potential 137
Positive current 138
Powder welds
  impressed current systems 102
  safety 102
Preinstallation inspection 38–40, 41–42
Preinstallation tank testing
  double–wall 44–46
  single–wall 42–44
Pressure gauge
  line pressure tests 73
  pipes 75
Pressure
  and temperature 43
Pressure relief valve
  pipes 75
  tanks 42–44
Pressure test
  double–wall tanks 45–46
  pipes 74
Pressurized Line Leak Detectors 125–126, 148
Pressurized piping 79–80, 148
Probability of detection 115
Probability of false alarm 115
Procedure
  Double–walled tank air test 45
  soap test 46
Pumps 73
Pumps
  inspection 46–47
Purging 84

R

RCRA, *See* Resource Conservation and Recovery Act
Recommended practices 4
Reduction 137
Reduction potentials 133
Reference cell 139
References 5
Regulated substance 11, 12–14
Regulations 6
Release detection 115–116
Relining 17
Removal procedures 83
Repairs
  minor 41–42
Requirements
  tank design 14–16

corrosion protection 145
EPA 14–21, 143–146
leak detection 115–116
piping 20–21
spill and overfill protection 19, 115
Requirements of Subtitle I 149
Resource Conservation and Recovery Act (of 1984) 2, 9
Restraint
flotation 58–59
Return current
in impressed current systems 99
and batteries 136
Runoff 24
Rust 92, 133, 142–143

S

Sacrificial anodes 17, 95
installation 96
Sacrificial Anode/Cathodic Protection
installation 95
figure *140*
and corrosion cells 139
Safety 81
entering tanks 85
hold–down straps 64
purging and inerting 85
Safety Letter 81, 85
Salt bridge 137
Sand 51–52
SARA, *See* Superfund Amendments and Reauthorization Act
Shipping plugs 42–44
plastic 45–46
Single-walled tanks 42–44
Site Assessment 88
closure in place 90
Slab at grade 59
Sludge 81
closure in place 89
Soap test
tanks 43, 46
piping network 75
Soil conditions 116
Soil samples
site assessment 88
Soil vapor monitoring 117
Soils
stability 24–25
type 24, 24–25
Space requirements 24
Specific gravity 34
Spill containment 7
Spill and overfill containment
existing USTs 19, 115
piping 20–21
requirements 147
Spill and Overfill Protection 147
Spills 1
Stage I vapor recovery 75
installation 78
Stage II vapor recovery 75, *78*
Standard
minimum, piping 73
State fund 149
State programs 8
State required mechanism 149
Statistical inventory reconciliation 116, 117
Steel
and corrosion 93
Steel piping 97
Steel tanks
anchoring 64
Step–down transformer 101
STI–P3 design *141*
"Sticking" tanks 118
Stray current corrosion *94*
Stray electrical currents 100
Structural damage
hold–down straps 62–63

Submersible pumps 72, 73
Subtitle I 2
Suction lines
  installation 72
Suction stubs 72
Sumps 30
Superfund Amendments and Reauthorization Act 9, 149
Supplemental anchorage 58–59
Support 49
Surface runoff 53, 59
Swing joints 74
Symptoms
  petroleum toxicity 82

T

Tank atmosphere 83
Tank access 79
Tank coatings
  corrosion and 95
Tank damage 30
Tank failures 49–50
  common causes 49, 50
Tank reliners 81
Tank removal 7
Tank tightness tests 7
  leak detection 116
  requirements 150
Tanks *62*
  damage 30
  failures 49
  flotation 58–59
  handling 37
  improper handling 38
  large diameter 58
  manifolding 50–52
Tank-to-soil potential 98
Technical Standards 91
Temperature
  pressure tests 43–44
Temperature changes
  in pressure tests 43–44
Test samples 88
Test wires 40, 96
Tightness tests
  pipes 74–75
Toxicity
  petroleum 82
Traffic requirements 24
Transportation
  used USTs 87
Trench depth 69
Trenches 69–71
  spacing 69
Turnbuckle assembly *62*
Turnbuckles
  metallic 64
Two–piece Straps *63*
Types of release detection 116–125, *119*
Typical automatic tank gauge system *119*
Typical hydrostatic monitor assembly *122*

U

Underground petroleum storage tank systems
  number of 1
Understanding corrosion processes 133
Unions
  steel 73
Unstable soils 55
Used tanks 86
UST
  leaking 1
  installations 23
  premature failures 67
  temporary storage 40

V

Vapor monitoring 116, 147, 149
Vapor recovery equipment
  inspection 46–47
  installation 77
Vapors
  recovery 76
Vents
  building mounted 77, 77
Vent lines 75
  inspection 46–47
  installation 77
Voids 49–50
Volume measurements 120, 121

W

Water table
  depth 24–25
Well bores 125
Wells
  figure *124*
  installation 125
  monitoring 123
  observation 123
Wet installation 58–59
Workplace regulations 51–52

X

Xylenes 83

Z

zinc anodes 96